Redefining Value Where People and AI Meet

Measure Smarter.
Avoid the Balance Sheet Black Hole.
Establish What Intelligence is Really Worth.

Sundara Raman | Dr Jayanthi Raman

Redefining Value Where People and AI Meet: Measure Smarter. Avoid the Balance Sheet Black Hole. Establish What Intelligence is Really Worth.

ISBN: 978-1-7642731-9-0 : Hard cover

NATIONAL LIBRARY OF AUSTRALIA

A catalogue record for this book is available from the National Library of Australia

Disclaimer

While the publisher and authors have used their best efforts in preparing this publication, they make no representations or warranties with respect to the accuracy, applicability or completeness of the contents of this publication and specifically disclaim any implied warranties of merchantability or fitness for a particular purpose. No warranty may be created or extended by sales representatives or written sales materials.

This publication is for educational and informational purposes only regarding the subject matter covered and does not represent accounting, investment, financial, legal, medical, or professional advice. The advice and strategies contained herein are not intended to provide specific guidance for particular circumstances and they should not be relied on as the basis for any decision to take action or not take action on any matter which they cover. Readers should exercise their own skill and care in relation to any material contained in this publication and obtain professional advice where appropriate. To the maximum extent permitted by law, the publisher and authors disclaim all responsibility and liability, including any harm, claims, loss of profit, any incidental, commercial, consequential, or other damages, to any person or entity, arising directly or indirectly from any person or entity taking or not taking action from following the guidelines, information, instructions and recommendations in this publication.

For general information and contact details please visit our web site:
www.abouthumanai.net

Cover design: Lavami
Cover image: Licensed from © iStock.com/giemgiem

Contents

Part IV – How data analytics (AI/ML) drives business value

Part V – Evolution of ICT and data analytics

Part VI – Challenges in value measurement approaches

Part VII – Human vs automation with AI – An interdisciplinary framework

Table of Figures

Index of Tables

List of Abbreviations

Abbreviation	Description
A2A	Agent to Agent
AI	Artificial Intelligence
AI/ML	Artificial Intelligence/Machine Learning
AML	Anti-Money Laundering
AML/CTF	Anti-Money Laundering and Counter-Terrorism Financing
API	Application Programming Interfaces
APRA	Australian Prudential Regulation Authority
ATM	Automated Teller Machines
AUC	Area Under Curve
AUSTRAC	Australian Transaction Reports and Analysis Centre
BAU	Business As Usual
BCBS	Basel Committee on Banking Supervision
BI	Business Intelligence
BIS	Business Information Systems
BoM	Bill of Material
Capex	Capital Expenditure
CART	Classification and Regression
CDO	Chief Data Officer
CDR	Call Detail Records
CEO	Chief Executive Officer
CIO	Chief Information Officer
CMO	Chief Marketing Officer
COBOL	Common Business Oriented Language
CPA	Certified Practising Accountant
CPU	Central Processor Unit
CRISP-DM	Cross-Industry Standard Process for Data Mining
CRM	Customer Relationship Management
DAVA	Data Analytics Value Assessment
DAVE	Data Analytics Value Estimation
DAVF	Data Analytics Value Framework
DAVO	Data Analytics Value Optimisation
DAVR	Data Analytics Value Realisation
DP	Data Processing
DPI	Deep Packet Inspection
DS	Data Science
DSM	Diagnostic and Statistical Manual of Mental Disorders
DSS	Decision Support System

DVB-SI	DVB Service Information
DVB-T	Digital Video Broadcasting-Terrestrial
EAD	Exposure at Default
ECG	Electrocardiogram
EPG	Electronic Program Guide
ERP	Enterprise Resource Planning
ETL	Extract, Transform and Load Process
F&PM	Financial Accounting and Performance Management
FM	Financial Metrics
FORTRAN	Formula Translation
GAAP	Generally Accepted Accounting Principles
GDP	Gross Domestic Product
GE	General Electric
GenAI	Generative AI
GP	General Practitioner
HBR	Harvard Business Review
HITL	Human-in-the-loop
HR	Human Resource
HTML	Hyper Text Markup Language
IBM	International Business Machines
ICL	International Computers Limited
ICT	Information and Communication Technologies
IDF	Inverse Document Frequency
IE	Information Engineering
IFRS	International Financial Reporting Standards
IFTI	International Funds Transfer Instruction Reports
IoT	Internet of Things
IP	Internet Protocol
IRB	Internal Ratings Based
IRM	Information Resource Management
IRR	Internal Rate of Return
IS	Information Systems
IT	Information Technology
IVR	Interactive Voice Recorder
JSON	Java Script Object Notation
KNN	k nearest neighbour
KPI	Key Performance Indicator
LDA	Latent Dirichlet Allocation
LEA	Law Enforcement Agency
LGD	Loss Given Default
LLM	Large Language Model

MCP	Model Context Protocol
MIT	Massachusetts Institute of Technology
ML	Machine Learning
MPEG	Moving Picture Experts Group
MPP	Massively Parallel Processing
MTBF	Mean Time Between Failure
NLP	Natural Language Processing
NPS	Net Promoter Score
NPV	Net Present Value
OCR	Optical Character Recognition
OECD	Organisation for Economic Cooperation and Development
Opex	Operating Expenditure
OR	Operations Research
OTA	Over the Air
OTT	Over the Top
P&L	Profit and Loss
PD	Probability of Default
QoS	Quality Of Service
RAG	Retrieval Augmented Generation
RFID	Radio Frequency Identification
ROI	Return On Investment
RPA	Robotic Process Automation
RPD	Recognition-Primed Decision
RWA	Risk Weighted Assets
S&D	Sales and Distribution
SMP	Symmetric Multi-Processing
SMR	Suspicious Matter Reports
SMS	Short Message Service
SVP	Support Vector Machines
TCO	Total Cost Of Ownership
TF	Term Frequency
TFP	Total Factor Productivity
TPC	Transaction Processing Performance Council
TQM	Total Quality Management
TTR	Threshold Transaction Reports
VaR	Value At Risk
VDU	Visual Display Unit
WT	Weight
WWW	World Wide Web
XML	eXtensible Markup Language

About the Authors

Sundara Raman is an industry veteran with five decades of international experience in IT, information engineering, data science, management consulting, and solution value selling. He provides pragmatic strategies, and tools for measuring and tracking the long-term value of data analytics (AI/ML) investments with results delivering compelling ROI.

Dr Jayanthi Raman is an academic researcher and clinical psychologist, with decades of expertise in decoding how humans think, decide, and behave. She bridges the worlds of human and artificial intelligence, that will empower readers to see the real strengths and challenges of both human and machine intelligence offering clear, actionable insights in an AI-driven world.

Preface

Context for this book

Telenor is an international telecommunication company with a history that spans over 170 years serving over 200 million customers worldwide. During one of the business value consulting engagements with an operating company of Telenor Group, the Chief Executive Officer (CEO) stated his belief that the investments in enterprise data analytics generated huge value. However, visibility was limited to the cross-sell and up-sell revenue growth reported by the marketing team from analytics-driven campaigns. In the same company the Chief Data Officer (CDO) said that he thinks his data scientists and advanced analytics teams are a lot more valuable than the data analytics assets held by the company and he wanted to quantify the value of his teams.

On a different consulting assignment, the Chief Marketing Officer (CMO) of a multi-national service provider asked if it was possible to quantify the value of data analytics that leverage advanced analytics with Machine Learning (ML) capabilities.

In the above cases, the data analytics (AI/ML) business value audit during the consulting practice demonstrated more than 330% Return on Investments (ROI) measured by Internal Rate of Return (IRR). But this was not apparent in the enterprise performance reporting and the balance sheet. Subsequent consulting work on data analytics business value surprisingly indicated that there are many enterprises in the commercial and not-for-profit sectors that are in the same predicament.

Why value?

First and foremost, why focus on value? In the past as well as in recent times, business intelligence (BI) and data analytics (AI/ML) have been generally considered as an overhead and a necessary infrastructure cost for doing business. Because overhead costs are indirect business expenses – they are not directly tied to the products and services that generate value – they are bound to be under constant scrutiny during rounds of cost-cutting. As a result, the very investments in data analytics that are most needed to help organisations survive and contribute long-term value have been the most at-risk.

The realisation that data and information represent the *fingerprint* of their customers' interactions across all channels and product usage has led to a shift in emphasis on value. Thus, data can be tapped into using analytics to provide a unique source of competitive advantage.

However, as evidenced from the management consulting practice and discussions with the C-level executives of large enterprises, those organisations wanting to measure value of data and analytics find it difficult to do so due to a range of factors, including short falls in Generally Accepted Accounting Principles (GAAP), International Financial Reporting Standards (IFRS), national/local financial measurement practices and lack of well-established data analytics management and value governance.

The 'shining star' of data analytics values vanishes into the black hole!

There are a range of financial benefits resulting from data analytics (AI/ML) that can be quantified as cash flow improvement, cost avoidance, cost reduction, retained revenue, revenue growth, revenue recovery, margin improvement and profit improvements. These eventually show up on the *bottom line* and end up in the enterprise balance sheet as general reserves and/or retained earnings. In essence, the assets minus liabilities in the balance sheet are balanced by shareholder funds, general reserves and retained earnings as net of profit and loss.

However, the values generated by data analytics are neither directly attributed to the *top line* nor the *middle line*. ***Instead, they vanish into the bottom line without trace!***

This book deals with the importance of measuring, attributing and presenting data analytics value. It examines the challenges involved in this process and puts forward strategies to overcome them. These strategies are illustrated with real-world examples, case studies and use cases drawn from several industries.

Strategies and tactics for valuing data analytics (AI/ML)

Much of this book is a reflection of learning from and turning points in decades of international experience in business information systems (BIS), information engineering (IE), data science (DS), management consulting, business value engineering and value selling in banking, energy and utilities, manufacturing, retail and telecommunication industries.

This book also draws upon the research and a survey of over 100 senior executives of national and multi-national organisations in New Zealand and a survey of over 200 senior executives of national and multi-national organisations in Australia. The New Zealand survey asked the business and IT executives how they measured value of *information systems investments,* whereas the Australian survey asked the business and IT executives how they measured value of *data analytics investments*. Given that in most organisations IT department supplies and provides service to business functions and divisions across the enterprise as its internal customer, responses from business executives and IT executives were separately reviewed to consider each of their points of view in the survey.

This book is intended for anyone to learn from tried and tested methods and processes in data analytics value attribution, reporting and communication. They will also find foundational concepts in data, analytics and economic theories of value very useful.

Students, young professionals and mature data analytics practitioners alike will find the history of the evolution of data analytics and value engineering across the different eras of computing over five decades of experience not only interesting but also thought provoking.

Business analysts and data analytics professionals will find the topic on value optimisation approaches using data reuse techniques and tools very useful. They will also find the business outcome focused data analytics use cases and case studies from the different industries covered in this book very handy.

Data analytics project managers can make use of the approaches to value realisation and value auditing described in the respective chapters.

Data analytics strategy development teams, and investment oversight and audit teams will find the content on value estimation and data analytics road map development techniques as well as reference values across business functions and divisions in their respective industries as useful tools. They will also find the know-how on sweating data analytics assets to grow value very useful.

Sales or Sales Support professionals who struggle to sell the unique features and functions of solutions to their prospects and customers will find the tools and techniques in this book a refreshing way of articulating measurable value to differentiate their software solutions.

Additionally, the readers will also benefit from the insights provided by Dr Jayanthi Raman, an academic researcher with decades of experience as a practising clinical psychologist with a professional background in understanding human cognition and the biases that shape information processing. She evaluates the strengths and limitations of the human mind and the machines, along with recommendations for dealing with the dilemma of investing in human and/or artificial intelligence.

All readers will find the interdisciplinary framework useful that evaluates job task characteristics, skill requirements, cognitive biases and limitations, and offers strategy for automation with artificial intelligence (e.g. Agentic AI) and/or augmentation with human-in-the-loop (HITL). They will also find the topics on methods of aligning stakeholders' interests with perceptions of value useful.

The book is sprinkled with nuggets of knowledge gained by the authors on what worked, what did not work and why, which we hope will be invaluable for readers.

The scheme of this book

Part I covers the background surrounding the problems with data analytics value measurement and its importance for organisations. It then introduces the scope of this book, definitions of words, phrases and key terms used.

Part II explores the concepts and perceptions of value and explains which of the economic theories of value applies to data analytics.

Part III develops the concepts of value chain in the organisation, measures of value metrics and data analytics value drivers.

Part IV describes how data analytics drives value and provides methods and processes for measuring and attributing its value.

Part V reviews history of information technology industry trends, and evolution of data analytics, value engineering and stakeholders.

Part VI follows directly from earlier parts of this book and analyses data analytics value challenges. Part VI develops data analytics value foundation to address many of the challenges.

Part VII contains an evaluation of the strengths and limitations of the human mind and the machines and provides recommendations and a strategy for dealing with the dilemma of investing in human or artificial intelligence (e.g. *Agentic AI)* to support judgement and decision-making with the objective of optimising and attributing their value contributions to business outcome.

Part VIII builds on strategic and tactical approaches in data analytics value measurement, considers human and/or *Agentic AI* decision-making and offers an interdisciplinary framework that blends business, cognitive psychology, data science and governance with methods, techniques, tools and benchmarks for effective management of value.

Part IX develops advanced concepts dealing with data analytics value governance and capability ranking with illustrative examples.

Part X contains case studies by referencing financial services and telecom industries.

Part XI includes concluding thoughts.

Acknowledgements

We gratefully acknowledge the contributions of our colleagues, friends, data analytics and business value professionals who supported the writing of this book. The content presented here draws on many years of professional practice in data analytics, consulting, and valuation, as well as decades of experience in clinical and cognitive psychology.

We acknowledge with gratitude the contributions of the scientific, economic, analytical, psychological, academic communities, and intergovernmental organisations, whose research efforts have been instrumental in informing this book and in identifying emerging trends.

We would also like to acknowledge the many colleagues, friends, and clients whose direct and indirect contributions, collaboration, and shared experiences over the years have shaped our thinking and enriched this work.

Our special thanks to Delphian Books for their diligence and for making the publication of this book possible.

Finally, we extend our deepest gratitude to our families and children for their unwavering support, patience, and encouragement throughout the writing process.

We have endeavoured to make this book as comprehensive, accurate, pragmatic, and engaging as possible. Ultimately, what matters most to us is the perspective of our readers. We warmly welcome your feedback and reflections and invite you to share your thoughts with us via our website.

www.abouthumanai.net

Sundara Raman and Dr Jayanthi Raman
Sydney Australia
April 2026

Part I

Introduction

Part I of this book provides background surrounding the problems with data analytics value measurement and its importance for organisations. It introduces clear definitions of words, phrases and key terms used, their meaning as well as scope of this book, and aims to overcome ambiguities surrounding widespread notions of data, information, analytics, Artificial Intelligence (AI) and their measurement of value.

CHAPTER 1

Background and scope of this book

Overview: This chapter provides background encompassing the core issues in data analytics value measurement and discusses why it is important for organisations to address these.

This chapter also aims to overcome ambiguities surrounding definitions of data, information, analytics, AI and measurement of value caused by the explosion of data following digital revolution. In the wake of the emerging data economy as a result, this chapter clarifies the scope of this book with a focus on providing methods, approaches and tools for measuring long-term value of data analytics in the microeconomy.

Data analytics (AI/ML) value attribution problem

Consider the two examples below and see for yourself where the value of data analytics went and how the value created was attributed in your organisation.

In a deregulated energy and utilities market, the price of the highest bid in the wholesale energy market required to meet demand dictates the 30-minute clearing price. Energy retailers have recognised that the traditional quarterly meter readings do not provide the ability to forecast usage and demand when the volatile spot prices change every 30 minutes, potentially leading to an unprofitable outcome. Armed with granular Interval Meter usage data, together with corresponding weather data, energy retailers are able to develop accurate predictive modelling that enables them to be competitive and profitable in a free market.

Here is another example. Many organisations do not publicly admit that they are losing revenue due to fraudulent activities but it is systematically revealed in internal audits with up to 5% of annual revenue being lost. For an organisation with a billion dollars in annual revenue, the loss can be $50 million. Even if only 10% of this is recovered, it translates to $5 million *top line* and *bottom-line* improvement. This is achieved with only a small investment in analytics applied on the data the organisation already owns. Anti-money laundering, identity theft, SIM Box and IDD Premium Rate numbers are all known examples of fraud that require advanced analytics to detect ever-evolving patterns of fraudulent activities, and to take preventive actions in time, thereby creating value for the enterprises.

In the examples above you may observe that an initial business case was established based on the cost benefits of enhancing business processes through data analytics compared to previous processes. However, as the organisation continues to reap ongoing benefits from these investments, tracking the value becomes increasingly challenging. This complexity arises from the continually shifting data analytics landscape within the organisation, compounded by the growing volume and variety of data. Also, the introduction of various analytic techniques and algorithms through IT projects and processes – such as DevOps, DataOps, ModelOps, AnalyticsOps, and MLOps – can obscure the benefits outlined in the initial business cases.

Why quantify and attribute data analytics (AI/ML) value?

In general, investments in a channel correlates with the channel's ability to generate and contribute towards achieving the sales target. So, any commission from sales achievement can be appropriately credited to the channel that was instrumental for the sales revenue. If the channel is not instrumental in achieving sales revenue goals, then the investment in that channel is futile.

In a multi-channel strategy, data analytics helps to track customers' product searches and buying behaviours across different channels that influence purchase intentions and buying decisions. This enables the marketing team to determine which channels influence achievement of their respective sales revenue target, so the value contributions of the channels can be effectively tracked and credited appropriately.

Without data analytics, the retail outlet where the customer finally bought the product/service usually earns most sales commission, even though the retail outlet may not have influenced the customer's final decision to buy! The customer's buying decision may have been influenced by recommendations from a social network channel rather than the retail outlet which served as a convenient location for the customer to purchase. Recognition of revenue and commissions to a wrong channel that did not influence the sale does not bode well for marketing spend in those channels!

The application of data analytics for customers buying behavioural prediction using AI/ML enables marketing executives an optimum investment strategy for channel distribution network with an appropriate revenue contribution and associated sales commission models. It also allows the ability to quantify and attribute the business value of data analytics that drives such important decisions.

In a similar vein, business value of data analytics can be quantified, attributed and traced for all data analytics initiatives using the framework, methods and tools described in this book.

Data analytics (AI/ML) are corporate assets

Treatment of data as an asset in the balance sheet is also fraught with difficulties. Although data and information continue to provide growing current and future economic value to the enterprise, **such data is never captured as either tangible or intangible assets**. Instead, the hardware and software that make up the bulk of the initial investment end up as tangible/intangible assets with a *book value* bearing the purchase cost. This is then subject to amortisation/depreciation (i.e. diminishing value).

Data scientists and analysts who provide the smart and advanced analytics skills as well as subject matter experts, managers and decision-makers in organisations are not treated as assets. Instead, they are considered as costs in the Human Resources (HR) system that roll into profit and loss statements as '*Labour costs*', '*Staff costs*', '*Salary expenses*'. Technology projects and data analytics projects are either capitalised and appear as tangible/intangible assets and amortised/depreciated costs or written off as expense, so disappear into profit and loss statements as '*Information Technology costs*' along with other expenses.

The following three examples from banking and telecommunications industries illustrate these points.

Table 1 is a snapshot of the published annual financial report (year 2025[1]) of the Commonwealth Bank of Australia, a bank with annual revenue of $28.29 billion and operating expenses of $13.01 billion. It has reported staff costs as 61% of total operating expenses and IT costs of 18% of total operating expenses in the profit and loss statement. This highlights the employee-centric nature of organisations that continue to rely on human cognition, talent and skills. Data analytics costs are not reported separately and are assumed to be part of IT costs. While employees and data analytics are not reported as assets in the balance sheet, computer software is reported as intangible assets with net amortised costs – a small fraction (i.e. 0.2%) of total assets.

1 https://www.commbank.com.au/content/dam/commbank-assets/investors/docs/results/fy25/2025-annual-report.pdf

Table 1: Snapshot financial statement of a large bank

Profit and Loss statement – Revenue and expenses		Remarks
Total Net Operating Income	$28.29 billion	
Total Operating Expenses	$13.01 billion	
Employee Expenses (i.e. Staff Costs)	$7.97 billion	Employee expense represents 28% of Revenue. **61% of Operating Expenses**
Information Technology (IT) Expenses	$2.38 billion	IT expense represents 8% of Revenue. **18% of Operating Expenses**
Data Analytics Expenses	Not reported	Data analytics expenses are not reported separately but are assumed to be included in IT expenses.
Balance sheet – Assets		**Remarks**
Total Assets	$1,353 billion	
Computer software – Intangible Assets at Amortised Costs	$2.60 billion	Computer software as intangible assets represents 0.2% of total assets
Employees	Not reported	Employees are not reported as assets
Data analytics	Not reported	Data analytics are not reported as assets

Similarly, Table 2 contains parts of the published annual report (year 2025[2]) of Telstra Group Limited, an Australian telecommunication company with annual revenue of $23.61 billion. It reported human resources as labour costs with 26.6% of total operating expenses and no separate data analytics or IT costs in the profit and loss statement. Software is reported as intangible assets with net of amortised costs of 10.4% of total assets (i.e. $44.96 billion), while the value of data analytics is not reported in the balance sheet.

2 https://www.telstra.com.au/content/dam/tcom/about-us/investors/pdf-g/telstra-annual-report-2025-interactive.pdf

Table 2: Snapshot financial statement of a large telecom

Profit and Loss statement – Revenue and expenses		**Remarks**
Total Income	$23.61 billion	
Total Operating Expenses	$14.98 billion	
Employee Expenses (i.e. Staff Costs)	$3.99 billion	Employee expense represents 16.9% of Revenue. **26.6% of Operating Expenses**
Information Technology (IT) Expenses	Not reported	IT expenses are not reported separately but are assumed to be included in 'Other expenses'.
Data Analytics Expenses	Not reported	Data analytics expenses are not reported separately but are assumed to be included in IT expenses in 'Other expenses'.
Balance sheet – Assets		**Remarks**
Total Assets	$44.96 billion	
Computer software – Intangible Assets at Amortised Costs	$4.69 billion	Computer software as intangible assets represents 10.4% of total assets
Employees	Not reported	Employees are not reported as assets
Data analytics	Not reported	Data analytics are not reported as assets

As a further example, a tier-1 bank operating in Australia and New Zealand undertook several initiatives: they closed down many of its branches in order to reduce costs by replacing employees with Automated Teller Machines (ATMs), and they invested in IT and data analytics to support automation in customer management. In response to this, the bank received complaints from customers who preferred fact-to-face dealings, and relationships with bank employees in their day-to-day transactions and interactions. When many of its customers switched to other banks offering better customer services, the bank reversed its decisions to reopen its branches and enhanced its portfolio of personalised products and services. This underscores the value of unique skills and capabilities provided by human resources in comparison to technology alone.

Crude oil or fossil fuel or renewable resource

When Big Data began to take root, the IT industry used the analogy of crude oil reserves to describe its benefits. This notion of Big Data as the mineral resource assumes that data is a natural resource waiting to be exploited. Inherent in this concept is the fact that the natural resource has a known up-front value and is a depleting asset that diminishes in value over time. Unlike the mineral resources, data and analytics do not have a preset value and are not a depleting resource. Instead, they are a renewable resource with increasing value over time. Big Data is the cumulative effect of continual increments of an organisation's internally generated data and externally sourced data. It is a renewable growth asset that propagates in volume and variety in the corporate data repository (the *data lakehouse*). It is driven by competitive pressures and innovation in commercial enterprises, and all competing for scarce funds in not-for-profit sectors. An organisation's data lakehouse can be enriched with analytics to generate actionable insights that differentiate it from the competition. This drives unique and greater ongoing long-term business value.

And no two data lakehouses are alike! As a consequence, each organisation takes its own unique journey in data analytics with varying levels of capability and value outcome as a result.

Gaps in GAAP

Accounting standards such as Generally Accepted Accounting Principles (GAAP) and International Financial Reporting Standards (IFRS) do not help either and are not conducive to allow for value recognition and valuation of data analytics (e.g. difficulties in assigning value of human talent in the accounting books and shortcomings in valuing enterprise data as either tangible or intangible assets). Consequently, the balance sheet does not provide direct visibility of the data analytics assets, thereby blindsiding the C-level executives of its true value in investment decisions.

Nevertheless, every organisation has a well-oiled machinery that governs and effectively manages talent acquisition, periodic human resource performance and incentives to account for the value of its people. But not many organisations can pride themselves in successfully governing and managing data analytics value! The data analytics value framework and governance included in this book will provide indispensable resources for such organisations.

Digital revolution: Data economy versus economic value of data analytics

Back in 2011, the author's attention was attracted by an article in *Time* magazine: '*YOUR DATA FOR SALE: Everything about you is being tracked – get over it. By Joel Stein*'.[3] Stein describes the enormous potential and growth of the multi-billion-dollar industry based on the collection and sale of personal and behavioural data. This data was mined using statistical techniques and algorithms that were being refined more and more every month to support the targeted advertising. Recently, a LinkedIn report mentioned that the US raised nearly half a trillion dollars in private investment in AI, higher than the rest of the world combined from 2013 to 2024, with 44.5% of investments in just two years between 2023 and 2024.

In related news, in 2016 *Harvard Business Review* published 'The Simple Economics of Machine Intelligence' claiming that the year 1995 defined the start of the *Digital era* – a new economy with key capabilities of search and communication made easier for a lower cost. This era marked the emergence of Machine Learning (ML) and was expected to result in a decrease in the cost of predictions, which would lead to reduction in cost of goods and services in industries.

Not surprisingly, the emergence of the digital economy has attracted the attention of the offices of national statistics and bureau of economics who see opportunities for enormous growth and impact of this in the macroeconomy. For instance, 'National Bureau of Economic Research'[4] provides a framework for measuring data assets as conceptually encompassed in intangible capital and concludes that 50% of intangible capital is *data capital*. This is created by processing and transforming digitised information into usable knowledge in an economy. This report states that business spending on data accumulation and transformation, and spending on the conduct of data analytics are intangible capital investments.

The kind of companies mentioned in the *Time* magazine article by Joel Stein that have large intangible assets as market capital are known in the industry as *BigTech* companies. The business models of these companies are focused on data monetisation strategy, selling data and information in the market as non-physical products and services known as *data products*. Naturally, the national bureau of census and statistics offices are viewing this as a social and economic resource, and expect this to play an important role in the macroeconomy over time by impacting the Gross Domestic Product (GDP), per-capita income, etc. They expect this

3 https://content.time.com/time/covers/0,16641,20110321,00.html

4 Francesco Bontadini, Carol Corrado, Ilaria Iommi, and Chiara Jona-Lasinio, 'Data, Intangible Capital, and Productivity,' National Bureau of Economic Research, June 7, 2024, https://www.nber.org/system/files/chapters/c14737/c14737.pdf

to provide a solid evidence base for national government policy-making. (While BigTech companies can be credited for their disruptive innovations in the economy, this book is neither concerned with data analytics value for companies with a focus on data monetisation nor its contributions to the macroeconomy.)

This book is concerned with firm level data analytics value creation and management in the microeconomy, where thousands of for-profit and not-for-profit organisations operate. These organisations aspire to be data analytics driven organisations. They are concerned with providing superior products and services to their customers, and a decent return on huge investments made by their stakeholders. By taking advantage of internal data generated in the process of running their operation and by applying advanced analytics (AI/ML) algorithms and techniques on the data, these organisations are concerned with making strategic and tactical decisions on how to differentiate and compete in their respective industries. The internally generated data is often supplemented with external data, such as customer credit rating data from credit bureaus and census data from national bureau of statistics to support decision-making in business operations.

Many of these firms in the microeconomy are capital intensive organisations that provide essential services to homes and businesses and operate under strict national and state regulations and consumer regulations. For instance, a telecom service provider has invested $35 billion, which is accounted for as 'Property, plant and equipment' in the balance sheet. This investment represents the fibre broadband communication infrastructure with connection that serves over 8.6 million homes and businesses. This infrastructure invested and commissioned by the telecom network and service provider also acts as the *plumbing* for the BigTechs without which the BigTechs would not be able to provide their *Over the Top* (i.e. OTT) data products and services in the economy. OTT is the term used in the telecom industry to denote how the BigTech firms deliver the products and services over the internet that ride on top of the telecom services providers' capital intensive infrastructure (such as fibre broadband, 4G/5G cell tower networks), rather than using their own dedicated network investments to connect to these homes and businesses!

Another example is that of an organisation that provides essential electricity transmission and distribution network serving 2 million customers with a regulatory fixed asset base of $19 billion.

Many such organisations described in the above examples make efforts to continually improve their business processes by taking advantage of data analytics to develop and support better products and services to their customers and, in the case of government organisations, its citizens. Case studies and use cases from many such organisations in different industries operating in the microeconomy are referred throughout this book to illustrate economic value of data analytics attributable to investments by these organisations.

Does benefit equate to value?

A few years ago, a colleague was reviewing a document dealing with data analytics entitled 'Business Value of Advanced Analytics'. Throughout the document the word 'benefits' was mentioned several times but the word 'value' was not mentioned at all, other than on the title page. Neither was there any method describing how to quantify the benefits other than qualitative descriptions of the benefits of advanced analytics solutions. While this document had its merits in describing the advantages of a particular method or solution when comparing with others, it becomes extremely difficult or impossible and increases the complexity when multitude of factors are considered in decision-making.

Generally, any set of metrics using mathematics, accounting and economic principles provides a reliable and efficient basis for developing 'value' on which decisions can be made. Nevertheless, the absence of a definition of value often leads to misinterpretation and is problematic due to a range of factors, including its subjectivity, perception and its dependence on people's skills, understanding and its usage. A separate chapter is devoted to it in Part II of this book.

For the sake of clarity, *value of data analytics* in this book refers to a set of quantitative metrics about how data and analytics are used for strategic and operational decision-making dealing with economic activities in organisations. It applies economic theories of value and they are measured by mathematical formulae using financial and accounting principles.

Definitions and meanings of data, information and analytics (AI/ML)

Similar to the varying understanding and meanings of value, data and information also have their fair share of ambiguity! When mentioning this book on data analytics value to a friend in Champaign-Urbana, Illinois in the US who is a specialist in library science, she jumped on the opportunity to help. She explained how the value of documents and publications in a library contributes to the economy by creating new knowledge by enriching valuable insights. In a similar vein, a friend in India who is a Certified Practising Accountant (CPA) remarked that his specialisation in information resource management (IRM) that deals with planning, budgeting and accounting will be helpful in our endeavour. We encountered similar reactions with different perceptions and meanings associated with data and information in several other disciplines.

In this book, *data analytics value* is focused primarily on the discipline of information and communication technologies, and information systems dealing with electronic processing of digital data for economic activities in organisations. Secondarily, this book deals with integration of data analytics and information systems with human judgement and decision-making in organisations to deliver business outcome. Thirdly, it coherently combines elements of social and cognitive sciences in an organisational context which provides a sound basis for relating human skills, knowledge creation, perceptions, biases and decision-making.

In this book, *data analytics* refers to the application of analytics involving mathematical and statistical algorithms on data to provide insights to support judgement and decision-making by people in the organisation. In this book, it is also interchangeably referred to as **data analytics (AI/ML)** to denote the application of advanced statistical techniques using machine learning (ML) algorithms to make predictions based on known data and information. Predictive analytics is increasingly being used in conjunction with the evolving trends of conversational AI in applications within the organisation.

Artificial Intelligence (AI) applications mimic human intelligence by taking advantage of deep learning neural network algorithms to perform tasks that require intelligence which is often being learned rather than directly programmed. *Generative AI* (GenAI) and *Agentic AI* are types of AI application that enables creation and synthesis of new content based on pre-trained Large Language Models (LLM). In this book, examples of Agentic AI are used but, not at the exclusion of many other advanced analytical algorithms involving machine learning functions that are used by organisations for various business and industrial applications to enable multi-genre analytics. We cover use cases that take advantage of multi-genre analytics comprising classification and regression models, cluster analyses, geo-spatial analytics, natural language processing (NLP), sentiment analyses, topic modelling, and so on.

Part II

Concepts and perceptions of value

Part II of this book addresses the misconception and perceptions surrounding the meaning of value and provides clarity and approaches for dealing with cost, benefit and value measurements by including references from economic theories of value.

CHAPTER 2

Perceptions and principles of value measurement

Overview: This chapter addresses the perceptions surrounding the meaning of value and provides definitions with demonstration of approaches for dealing with cost, benefit and value measurements. Furthermore, this chapter compares the various well-known economic theories of value measurement and identifies the most appropriate and best fit value measurement principle applicable for data analytics among the many economic theories of value.

Cost savings versus value enrichment

'Cost' can be defined as a resource given up or relinquished to achieve an outcome, often in exchange for another resource or thing. Any savings as a result of avoiding costs or recovering costs is often interpreted as benefits and 'value' gained. Why do organisations care about value? There are business reasons and accounting reasons to justify investments. Does it really matter if we use cost savings to describe benefits instead of value enrichment? Consider the following cost-based and value-based approaches to justifying investments for speed camera programs.

Cost-based approach:

- The cost of conducting the mobile speed camera program in a year is $14.6 million.
- To reach break-even point, which is the minimum quantity of events or things where the total revenue collected equals total costs, the police must catch 70,000 offenders per year, each with an average $209 infringement fee.

Value-based approach:

- The associated speed warning and safety awareness campaign is expected to reduce deaths from 120 per year to 80 per year.
- The financial benefit is estimated to be $300 million from avoidance of cost per fatality of $7.5m, involving use of ambulance, pathologists, hospital, police, social welfare, etc.
- The real value is 40 human lives saved!

When a speed camera at the corner of Oxford and Crown streets in Sydney earned $1.5 million in fines on a single day, probably justifying benefits of the investments by multiple folds, one of the newspapers reported, 'The speed camera was so successful they stopped issuing fines.' It said, the government stopped enforcing the fines and started giving people warnings instead!

Which of the above approaches is better for determining the value benefits? The answer rests on what is the object of valuation, how is value defined and measured, and what makes up a logical and pragmatic method for valuation. Of course, any investment decisions must consider costs, benefits (i.e. rewards) and risk exposures. The following three examples in trading cars further illustrates components of costs, value and justifications for their decisions.

Example 1:

- Model: 2024 Brand new BMW Z4 – sDrive20i M Sport with a price tag of $122,000.
- Costs: Includes production costs, distribution and selling costs; taxes and other applicable costs.
- Value for the seller: Gross margin during change of hands.
- Value for the buyer and benefits justification: The car serves as a utility function for commuting.

Example 2:

- Model: 2022 Second hand BMW Z4 – sDrive20i M Sport with a price tag of $92,000.
- Costs: Includes buying and selling costs; depreciation costs; taxes and other applicable costs.
- Value for the seller: Gross margin during exchange of hands.
- Value for the buyer and benefits justification: The car serves as a utility function for commuting.

Example 3:

- Model: 1965 Ferrari 275 GTB/C Speciale with a price tag of $26,400,000.
- Costs: Includes buying and selling costs; depreciation costs; taxes and other applicable costs.
- Value for the seller: Gross margin during change of hands.
- Value for the buyer and benefits justification: Rarity; collector's item; prestige.

How do the above discussions relate to data analytics? Many organisations consider data analytics investments as an overhead cost. As an overhead, data analytics investment is viewed as a necessary cost but one that does not add value. Cost is an important component, and it is prudent to focus on efficiencies of cost management, while also taking measures to create value for the organisation as a growth asset.

Relating data analytics value to economic theories of value

In this section, we will extend the discussion by considering some of the well-known classical economic theories of value. This will help in setting the background for identifying appropriate treatment for measuring and managing data analytics value.

Intrinsic theory of value states that value does not exist as an object. It is the property of an object as against the perception or attribution of value determined by an individual.

Labour theory of value refers to economic value of a goods or services as determined by the total amount of necessary labour required to produce it.

Economic value is the value a person places on an economic good and/or service based on the benefit that they derive from the good. It is often estimated based on the person's willingness to pay for the good.

Exchange theory of value implies goods and services are subjective material use value and an objective change of hands value at the same time.

Utility theory of value is associated with price or value determined, based on how much 'use' or 'utility' individuals or groups receive from goods and services.

We previously reviewed emerging opportunities in the economy for data monetisation spearheaded by the innovation of BigTechs. In these cases, *value in exchange* or *exchange theory of value* will be relevant because many of the companies in the data economy operate their business by trading data and information as *data products.*

But, for the thousands of organisations in the microeconomy which are under considerations for this book, *utility theory of value* will be most relevant. *Value in exchange* is not useful for data analytics investments, as there is no exchange of hands in the daily operation of these organisations. But, in the event of mergers and acquisitions, data analytics may be considered as intellectual property for these organisations with *intrinsic value.* However, as discussed in the section Gaps in the GAAP, value of data analytics is not reflected in the balance sheet as a growth asset. Instead, it is often written off as expenses or depreciation and partially accounted for as a tangible depreciated asset or an intangible asset with amortised costs of acquisition. Value arises due to the ability of IT and data analytics to enhance the business performance of the enterprise[5] which is not separately tracked or reported in the accounting books.

'*You cannot manage what you don't measure*' – you may have heard that before! Certainly, this truism holds today. Choosing the right metrics is important because they influence behaviour of the entire organisational culture that affects management decisions and actions related to future strategies. This is stressed by John Hauser,[6] Professor of Marketing at MIT Sloan School of Management. If investments in data analytics are considered strategic, then a focus on attributing its value as a growth asset, rather than simply focusing on cost efficiencies, will help the organisation measure and manage data analytics effectively.

Much of the emphasis in the book is about treating data and information as growth assets by also considering a number of unique properties and characteristics of data, in comparison to physical assets. These are discussed in detail in the section in chapter 3 entitled 'Assets: Physical versus data analytics (AI/ML)'.

5 Marilyn M. Parker and Robert J. Benson, *Information Economics: Linking Business Performance to Information Technology*, Englewood Cliffs, NJ: Prentice Hall, 1988.

6 John Hauser and Gerald Katz, 'Metrics: You Are What You Measure!,' *European Management Journal* 16(5) (October 1998): pp 517-528.

Part III

Data analytics (AI/ML) value chain

In stark contrast to physical assets with their diminishing value, Part III describes the distinguishing characteristics of data and analytics assets with its growing value. It highlights the unique aspects of the data analytics ecosystem that has given rise to knowledge-based assets which have been instrumental in improving organisational processes and enriching decision-making capabilities in the enterprise.

Part III further describes the processes of value creation in an organisation by including the concept of traditional and digital value chains in economic activities. It also describes the opportunity afforded by the digital value chain for taking advantage of the enormous data generated in digital transformations. Following this, the need for managing the resources mix, such as labour, technologies and tools, in work processes that create value in the value chain is discussed along with tactics for guiding when to substitute and/or augment human tasks with technologies involving data analytics (AI/ML) and the emerging Agentic AI applications.

CHAPTER 3

Distinguishing characteristics of data analytics as a growth asset

Overview: In line with the central theme of this book to measure the long-term value of data analytics investments, this chapter provides detailed illustrations of the distinguishing characteristics of data and analytics as an asset with growing value as against physical assets that exhibit diminishing value. It also describes the unique aspects of the data analytics ecosystem that has given rise to knowledge-based assets (i.e. human capital) which are instrumental in improving organisational processes, decision-making and business performance that underscore its ongoing value contributions to the enterprise.

Assets: Physical versus data analytics (AI/ML)

It is important to define what constitutes data, how it is different from other physical resources and the unique characteristics of data that will assist in constructing and attributing the value contributions of data analytics (AI/ML).

Much like commodities, the availability and prevalence of data has led to improved products and services as well as disruptive innovations. However, one of its distinguishing characteristics is that the role of data has given rise to knowledge-based assets (also known as *Intellectual Capital*) in improving organisational processes and business models. Data has also contributed to enhancing productivity and economic growth.

The author has adapted[7] the descriptions of the unique characteristics of data from Organisation for Economic Co-operation and Development (OECD)[8] by including them within quotation marks in this section of the chapter. Each data characteristic is followed by illustrative narrations and examples from the author's industry experience which vividly encapsulate the unique characteristics of data.

7 This is an adaptation of an original work by the OECD. The opinions expressed and arguments employed in this adaptation should not be reported as representing the official views of the OECD or of its Member countries.

8 James Mitchell, David Ker, and Molly Lesher, 'Measuring the Economic Value of Data,' Going Digital Toolkit Note No. 20 (Paris: OECD, 2021), https://goingdigital.oecd.org/data/notes/No20_ToolkitNote_MeasuringtheValueofData.pdf.

- *'Data refers to recorded information in structured or unstructured formats, including text, images, sound, and video'.*"
    - Structured data refers to data in fixed formats arranged in row and column structure as in tables in printed documents or in Microsoft Excel spread sheets or in Relational Database tables.
    - Unstructured data refers to flexible texts in paragraph and sentence in documents or metadata about free format texts, images, sound and video that describe the formats of digital encoding of the free format texts, images, soundtrack and video in storage or streams. Software applications use the information contained in the metadata to decode the image, sound, video. Examples of free texts include Hyper Text Markup Language (HTML), eXtensible Markup Language (XML), JavaScript Object Notation (JSON). HTML also holds tags and attributes for images.
    - Moving Picture Experts Group (MPEG) provides a set of standards for storing and streaming audio and video by encoding, decoding, compression and decompression. For example, MPEG-I defines the standard for 'Coded representation of immersive media'. Part 28 of the standard describes 'Interchangeable scene-based media representations'. Here, it outlines various metadata for scene based immersive media. For instance, 'Lighting node' determines which objects in a scene are potentially visible to the camera used by the renderer for a scene; 'Camera node' defines the field of view, lens properties, and size of the view port that is output by a renderer of a scene.
    - Similarly, Digital Video Broadcasting-Terrestrial (DVB-T) provides a set of standards of which DVB Service Information (DVB-SI) and DVB-SI tables containing Electronic Program Guide (EPG) metadata are digitally transmitted over-the-air (OTA) and received through the antenna by the digital TV equipment for displaying on the TV screen in a semi-structured table format for the convenience of the viewer.

- *'Data is delineated from observable phenomena (i.e. the facts and situations whose attributes can be observed and recorded digitally, and that underpin the production of data). Two pieces of data are almost never identical as observable phenomena on which the data are based usually have unique attributes.'*
    - For example, when a patient browses the World Wide Web (i.e. the web) looking for booking an appointment with a General Practitioner (GP), the observable actions are

recorded as HTML data by the web server as a log (i.e. a weblog) of traceable events of interaction, including appointment booking and confirmation. After the appointment, an invoice transaction may be generated showing the fee for the GP consultation which will be recorded as invoicing event data. Following this a payment transaction may also be recorded as payment event data.

  - A similar process followed by a different patient at the same time or the same patient at a different time may also capture similar data events, but they are likely to be different and unique.
  - Another example is timetables of bus and train schedules and routes. This is data against which a department of transportation may record and track actual arrival and departure times as well as passenger load statistics and detours.
  - The production of data almost always occurs in operational systems as described in the above examples. Produced data from operational systems or transactional systems, also known as *source systems* in the language of the data analytics architecture, flows into a central repository in the data analytics ecosystem, which will be described shortly.

- *'Data can also arise often as a by-product of the standard production process.'*
  - For example, a manufacturing facility may continuously scan stages of product assembly on a conveyor belt and record all the events as data representing date, time, raw materials used (i.e. the Bill of Material – BoM), product quality, quantity, any defects detected and other related data.
  - Similarly, in the telecommunication industry, when a customer makes a call on their mobile smartphone, there are regular recordings of the date and time of call, calling party number, called party number, duration of call, the cell towers that connected the call at the call originating end as well as at the call terminating end, locations of the parties in the conversation and many other related data.
  - As with the telecommunication example above, the systems and/or equipment involved in the production process are also known as *source systems* in the language of the data analytics architecture. Data produced as by-products of the production process also flows into a central repository in the data analytics ecosystem.

- *'Data may have increasing or decreasing returns to scale. Sometimes, collecting more data can provide additional insights, but at other times more data adds little extra value (and likely some cost) but some organisations continue to accumulate and potentially hoard it.'*
  - Digitally generated online and web data in formats such as HTML, XML and JSON are considered very 'noisy' data due to massive volumes with a significantly small fraction of useful data. Massive volume is partly the result of the additional tags and other markups created by the HTML, XML and JSON standards. These tags and markups represent metadata which are overheads but are still required for use by the web servers and other applications which encode and decode to extract a relatively small number of observable and useful data relating to customers, patients and in many production processes, such as manufacturing, electricity generation, and telecommunication network signalling traffic.

- *'Data collection may have high up-front cost (i.e. investments in hardware, software, subscriptions to "Cloud" platforms and services and other related project costs) and incremental marginal cost. This can create barriers to beginning to collect data that would be useful or potentially financially lucrative.'*
  - Telecommunication service providers continuously record metadata of voice call, text message and mobile data traffic events in their network, known as *Control Plane* and *User Plane* signalling data. For a service provider with 10 million customers, the signalling traffic generated on the mobile network can add up to 100 terabytes of data per day. The huge capital expense involved in collecting, storing and processing this data creates a barrier to many telecom service providers. However, there is immense value in utilising such data. First, it provides insights on the service provider's customer behaviours, interactions and experience, creating a *fingerprint* of the customer. Secondly, this data enables several other benefits: it can be used to ensure high quality service by predicting and preventing network failures.

    For example, a global brand mobile service provider lost over a million customers in a matter of four months due to poor customer experiences, such as frequent dropped calls, poor smart phone upload/download speed as well as frequent loss of communication signal between cell towers. The service provider was unable to identify the root cause of the customers' problems as they did not collect the network signalling data, which could have provided such valuable information. As a result, they lost several million in revenue and the poor reputation in the market stunted revenue growth in acquiring new customers.

The new CEO and the management team quickly recognised the need for investing in collecting and storing the network signalling data. They invested in advanced analytics to process and predict customers' experience, which in turn enabled them to offer high quality service, recover from previous losses and to become profitable.

- *'Data use requires complementary and often on-going expenditures. Organisations may need to invest in complementary hardware and software to be able to process and analyse the data.'*

Figure 1 provides an overview of the complementary systems (i.e. platform, hardware, software, applications), business functions and users surrounding the data analytics ecosystem.

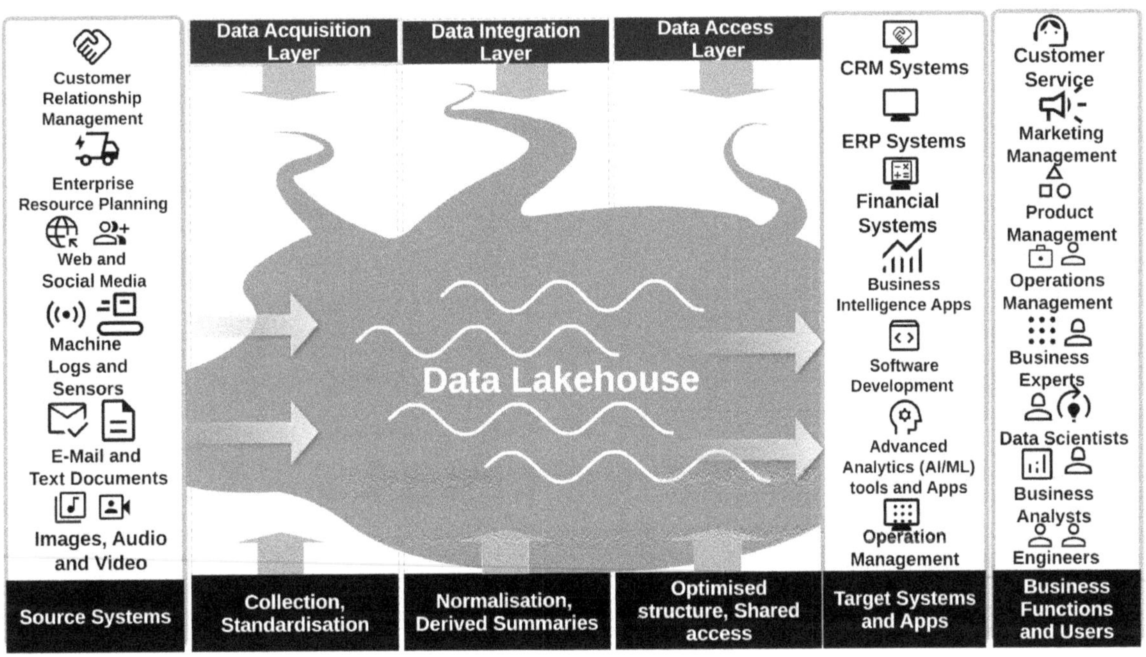

Figure 1: Data analytics ecosystem conceptual architecture

In a nutshell, the data analytics ecosystem integrates with a number of data sources on one side from operational systems and/or production processes in industries such as manufacturing, electricity/gas generation and distribution, telecommunication networks, aircraft operation besides other industries including retail, financial services, and on the other side it integrates with a number of destination operational systems, production processes, tools and applications.

Data analytics users (i.e. analysts, business experts and executives) from different business divisions and functions across an enterprise access these systems, applications and tools in the course of business activities for strategic, tactical, and operational decision-making.

- *'Two crucial aspects of data can affect its value: "Content" – the information the data embodies, and "Context" – the circumstances in which the data were gathered and are (or could be) held, analysed, and used. The context of collection will also affect the information content and quality of the data. For example, intended use may not be clear when collecting data or whether to collect data that has potential future use! It is hard to predict how the value of data might change.'.*

- *The next bullet point elaborates this with some examples.*

- *'Conventional analysis used to assess the value of physical assets and commodities cannot effectively capture the value of data as the production of data is often never fully "completed" (e.g. cycle of periodic data updates). Additionally, there may be a blurring of the line between the value created as a result of production and value stemming from the observable phenomena that underpin data products – (e.g. the value of the data asset can change with the addition of each newly recorded observable phenomena)'.*
    - For instance, the value perception of climate forecasts by weather bureau may have limited value for some but when a bank combines weather bureau's forecasts with its customers' mortgage loan data and physical location of the properties to determine credit risk, it may be more valuable in taking actions to prevent losses.
    - Similarly, in referring to an earlier bullet point, an Electronic Program Guide (EPG) that is broadcast over-the-air (OTA) to digital TV may have limited value but, when it is combined with viewers' video watching behaviour and viewers' demography, it may provide valuable information for targeted advertising for the service provider as well as the recommended video watch list for the viewer.
    - As we discussed previously, perception of value may be different between individuals or organisations – what is of value to someone may not be so for the other! Also, the actual value or perceived value can change over time.

- *Somewhat similar to the previous major bullet points, '... unlike physical resources, data are not susceptible to wear and tear. However, obsolescence can affect some data depending heavily on the context of use.'*

- Data, unlike traditional raw materials, accumulates, improves, and becomes more valuable as organisations refine it and extract insights. Unlike machinery that wears out, data becomes more potent with better algorithms and greater integration.
- For example, government regulators require organisations to retain several years of employee, customer and transaction data. In the case of income tax regulations in Australia, organisations are required to retain records for seven years, including transactions and business operations, employee details and financial records. Such historical data remains inactive and idle until situations call for their access, such as for providing evidence in a court of law. Much of the data collected and stored becomes obsolete without use, thereby incurring initial capital and ongoing expenses that can potentially diminish value.
- However, the case study example below illustrates the immense value of historical data in some contexts is illustrated in this case study. As required by regulation, a global telecommunication service provider has been retaining seven years of financial transactions and Call Detail Records (CDR) in the data analytics ecosystem that represent the metadata of all voice calls, short message service (SMS) and smartphone data usages by its customers. On one occasion the tax office claimed overdue taxes from this company on billed revenue from a period four years earlier. But the company encountered a problem with its billing system that originally calculated customers' network usage and generated the revenue reflected in the invoices sent to its customers. The billing system and financial reporting system did not contain enough details in their records to verify the tax office's claims. One of the data scientists in the company resorted to the historical data in the data analytics ecosystem that contained among much other data: the telecommunication network logs which, along with the billing and financial transactions, helped to verify, calibrate and reconcile the tax claims. They found that, instead of dues to the tax office, the company had overpaid tax to the tune of $300 million which was recovered from the tax office. In this context, the business value of investments in data analytics far outweighed the total costs incurred.

- *'Data about people have particular features. People create positive and negative incentives and impacts for others as they share information about themselves (e.g. through social networks or genetic analysis services) which may well be highly valued in principle by the individual. Behavioural information about a person in the context of products and services offered by an organisation may have some meaningful value to an organisation, but personal data are generally subject to additional legal controls, with compliance costs acting as another potential barrier to data collection.'*

- As mentioned earlier, data and information represent the *fingerprint* of an organisation's customers, product usage history and related details. This provides a unique competitive advantage to the organisation, making the data very valuable which, if lost, destroyed or stolen, would be impossible to replace and could lead to costly legal repercussions.
- Metadata which describes the data an organisation holds is also unique and provides an advantage due to the embedded meaning of data attached to them.

In summary, the distinguishing characteristics of data justifies in treating it as a long-term asset with growing value.

CHAPTER 4

Digital value chain in the enterprise

Overview: This chapter furnishes the process of how data analytics value is created in the organisation by making references to the construct of value chain in economic activities. It expands the value chain construct in the digital economy with illustrative examples of industry case studies by using business process engineering notations. It also highlights the opportunity offered by digital transformation for taking advantage of the enormous volume and varieties of data generated in the process and to grow business value.

Business value creation in the digital economy

One of the ways value is created in an organisation is to look at the economic activities involved in developing and delivering products and services. This can be achieved by involving the business divisions and functions of the entire organisation. This is best articulated by the *value chain* as described in Michael Porter's well-known book, *The Competitive Advantage: Creating and Sustaining Superior Performance.*[9] The value chain is a useful way of looking at how a company's business operations fit into the bigger picture as shown in Figure 2.

Digital transformation

How about a digital value chain? Digital computing technologies are used in economic activities by transforming traditional value chain models and profoundly altering conventional notions of how businesses are structured, how consumers obtain goods and services, and how regulators adapt to new regulatory challenges. Just by taking a sneak peek into the history of Radio Shack, as an example, reveals how the notion of digital has altered businesses!

Radio Shack, the digital electronic retailer, was once the go-to shop for innovators and engineers for a range of products. They filed for bankruptcy as they did not foresee the opportunity to transform into the new market dynamics. However, the new owners have taken advantage of digital transformation by using the stores as showcases for their products, as well as potential pickup and drop-off centres for online customers. Thereby they fulfilled the changing needs of customers by exploiting convergence of online and offline channels, often called omni-channel.

9 Porter, M.E. *The Competitive Advantage: Creating and Sustaining Superior Performance.* NY: Free Press, 1985.

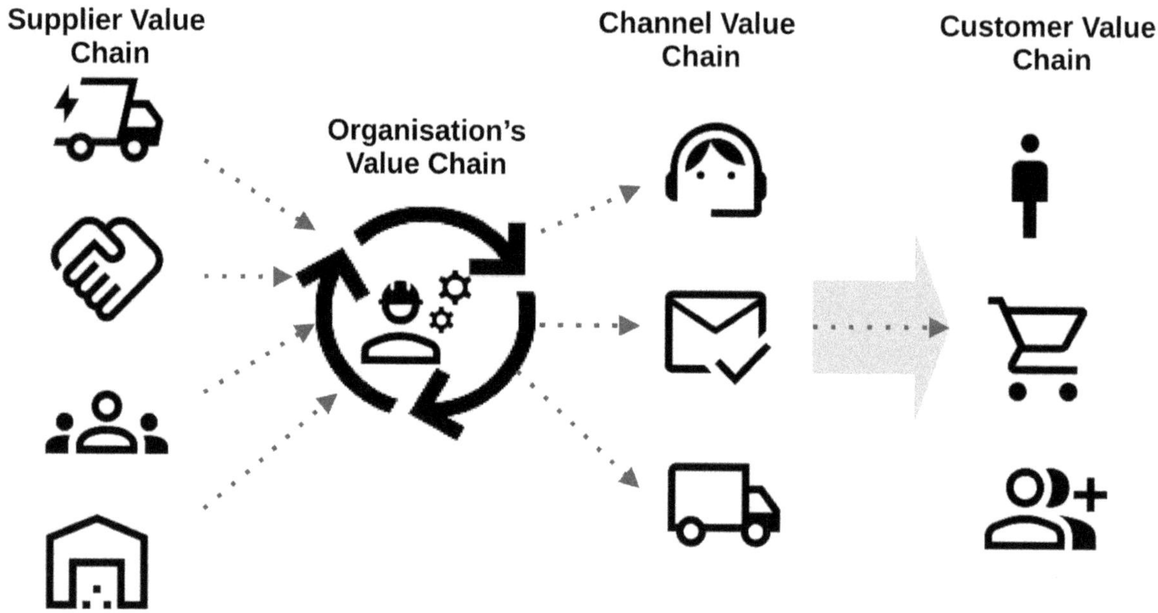

Figure 2: Value chain: The organisation's big picture (Adapted from Porter, M. E. *The Competitive Advantage: Creating and Sustaining Superior Performance.* NY: Free Press, 1985)

Beyond omni-channel marketing

Aligning business strategies to customer preferences and interrelated business processes is important. Digital transformation initiative is no exception to this because when done in isolation it can be counterproductive. For instance, as part of digital transformation, cutting channel costs by reducing retail outlets or contact centre seats often leads to the creation of low-cost channels, such as online self-service for bill payments. But, in case reviews, one service provider found that even after customers had paid their bills online they called the contact centre agent anyway, which lumbered the call centre with extra costs and defeated the original objective of exploiting low-cost online channels.

The reason why customers called the agent became clearer when analytics were performed on customer contact notes. This showed that these customers were anxious that their service was not mistakenly disconnected on assumptions of non-payment of bills and wanted confirmation from a real person that their payments were received. Armed with this new insight, the service provider modified the overall business process in their service engineering department to include an automated notification, along with an assurance of payment receipt to customers. The result was reduced calls to the contact centre agents. This change in business process also improved customer experience and satisfaction.

In essence, digital transformation is all about adapting and aligning an organisation's business processes to the changing attitudes and behaviours of customers. This can be achieved by leveraging the new use of digital technologies, such as the ubiquitous availability of broadband networks, connected smart devices and software applications (e.g. online portals, online chat, social networks, social media), that now reach over seven billion people worldwide, to communicate and interact with customers. However, digital transformation requires organisations to transform their business systems and employees' mindsets to take advantage of IT and data generated from business processes to drive value. Why is this important?

The answer can be illustrated through the experience of a senior retail distribution executive of a multinational telecom service provider who spoke about the opportunities that digital presented. While recognising the need to implement a digital strategy, they admitted that the end-to-end business process and the online systems were not integrated in their organisation – the Customer Relationship Management system (CRM) was not 'talking' to the Enterprise Resource Planning system (ERP).

It is essential that these two systems are integrated: when a customer places an online order to upgrade their smartphone and if they are promised same-day delivery, then the service provider needs accurate inventory and delivery logistics data. If these systems are not integrated, then neither accuracy of inventory levels nor commitment for same-day delivery could be guaranteed. This may lead to dissatisfied customers who are likely to turn to the telecom service provider's competitors. This underscores the importance of systems as well as data integration across the organisations' value chain. Digital transformation puts the data analytics ecosystem at the centre stage of the value chain.

Stickers are things of the past

The prospects of the paperless society have been long coming! But, with widespread availability of the digital infrastructure in recent years, it has now become an essential reality. Let's take a look at another example of *digital value chain*.

Commercial organisations are not the only ones taking advantage of digital transformation. One state government organisation in Australia has successfully transformed the vehicle transport industry across the country from a paper-based system into a full end-to-end digital electronic process by including the insurers, vehicle safety inspection service stations, payment processors and vehicle owners. No more fun flying paper planes though!

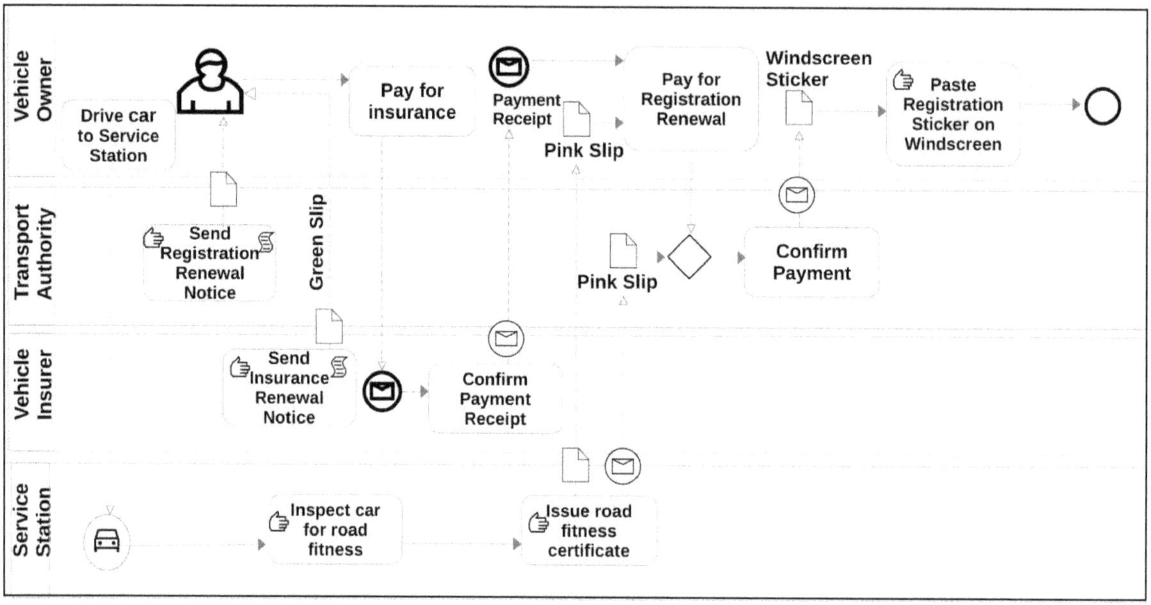

Figure 3: Previous paper-oriented vehicle registration and renewal process

Previous paper-oriented process

In the previous paper-oriented process described in Figure 3, registered vehicle owners received snail-mail documents from three different organisations each year:

- *Green Slip* – a new or renewal of compulsory third-party insurance policy from insurers
- Registration – a new or renewal notice of registration from the road transport authorities
- *Pink Slip* – a road fitness certificate from the service stations who perform vehicle safety checks

Once the authority received confirmation of the insurance and road fitness certification, they posted a new registration sticker to the vehicle owner. The vehicle owner then pasted the sticker label onto their windscreen, which was a regulatory requirement. Any violation of a new or renewal of registration found during random roadside inspections by police were notified to the vehicle owners by snail mail.

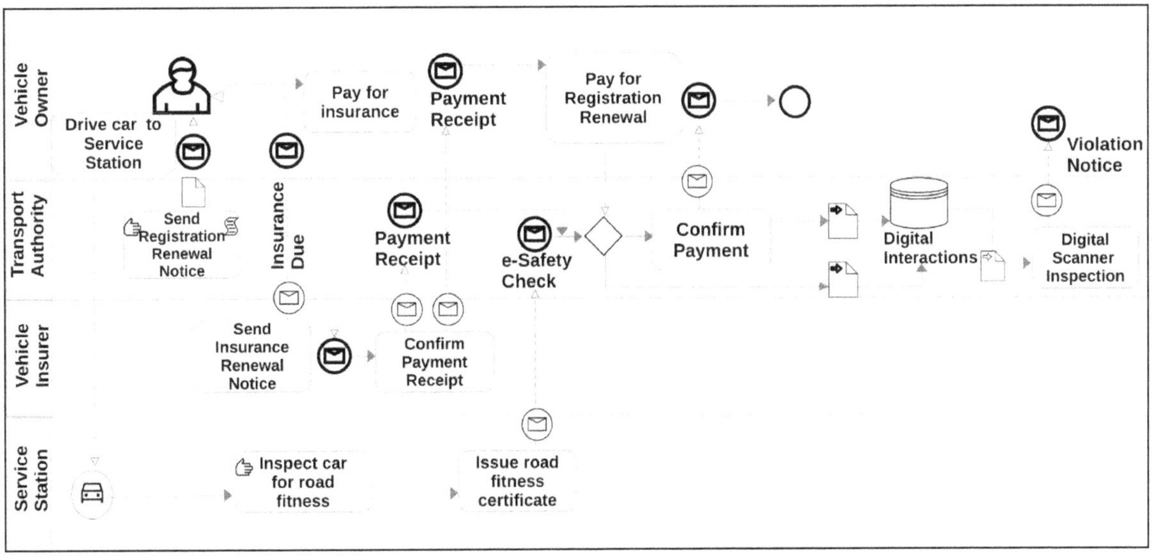

Figure 4: Paperless digitally transformed vehicle registration and renewal process

The paperless digitally transformed process

Now, in the digitally transformed process described in Figure 4 vehicle owners receive an email renewal reminder from insurers and online access from the road transport authorities. The insurance company sends a digital confirmation of the insurance policy payment directly to the road transport authority. Service stations also send digital confirmation of vehicle's road fitness (i.e. an e-Safety Check) directly to the authorities. In turn, the road transport authorities update the digital vehicle owner's registration details (which can be accessed, securely, by the owner), doing away with the registration sticker. Registration compliance is monitored by police officers using portable wireless scanners to scan vehicle number plates.

This way, the government organisation has enabled cost and time savings, and operational efficiencies across the full value chain involving all the partners and vehicle owners by integrating the various systems as well as critical data in their data repository.

CHAPTER 5

Value drivers in the enterprise

Overview: This chapter extends the discussions from the previous chapter and identifies value drivers and establishes the need for managing the resources, such as labour, technologies and tools, employed in work processes that drive enterprise value in the value chain. This leads to the discussion on approaches for organising the right resources mix to optimise business value. It requires a review of work situations and resource characteristics that can guide decision-making on when to substitute and/or augment human tasks with technologies in general, data analytics (AI/ML) and the emerging *Agentic AI* applications in particular.

Data analytics (AI/ML) value drivers

The digital value chain has also led to the creation of new types of data – such as unstructured data from email communication from partner networks and image data from number plate scanners – which were previously absent in the enterprise data analytics ecosystem. Analytics helps these organisations gain new insights from these new categories of data which can enable ongoing operational efficiencies from process improvements across the full value chain, thereby reducing cost of operation while improving customer/citizen experience and creating business value.

Increasing the business value of data analytics in the digital value chain requires managing the resources in work processes that create value. These resources include labour, technology, tools, technical skills, intuitive judgement, innovation, creativity and organisational policies to guide decision-making by determining which resources to invest in that will drive business value. However, many categories of technologies and tools are typically deployed which complement human workers in business operations that generate value by enhancing productivity and efficiencies which makes it difficult to determine value share of data analytics investments. Figure 5 contains examples of the different categories of resources used by organisations determined by the mix of resources required to orchestrate business processes to achieve business outcome.

In order to identify data analytics value among other technologies and tools, methods and processes are required to isolate and attribute value contributions of data analytics which we will discuss detail in Chapter 7 'Principles of data analytics value attribution'.

Historically, we can see how the technology and tools mix in work processes has evolved over the different eras. For instance, during much of Industrial Era, work processes involved physical capital and human labour as the basis of wealth and value creation in organisations. These included tractors in agriculture, and excavators and cranes in construction industries that complemented human workloads. In the post-Industrial Era, electronics played a significant role in augmenting as well as substituting human tasks, thereby enriching the value created in the organisation. These include robotics in assembly lines in the manufacturing industry, and X-ray, Ultrasound, and Electrocardiogram (ECG) machines in the healthcare industry.

In the Information Era, enterprises compete based on business processes by orchestrating and optimising resources consisting of people (i.e. knowledge and skills), IT and data analytics tools as factors of production, business outcome and organisational performance (see Figure 5). This has been growing in importance in the service sector where human resources has been most important in the organisational activities and tasks.

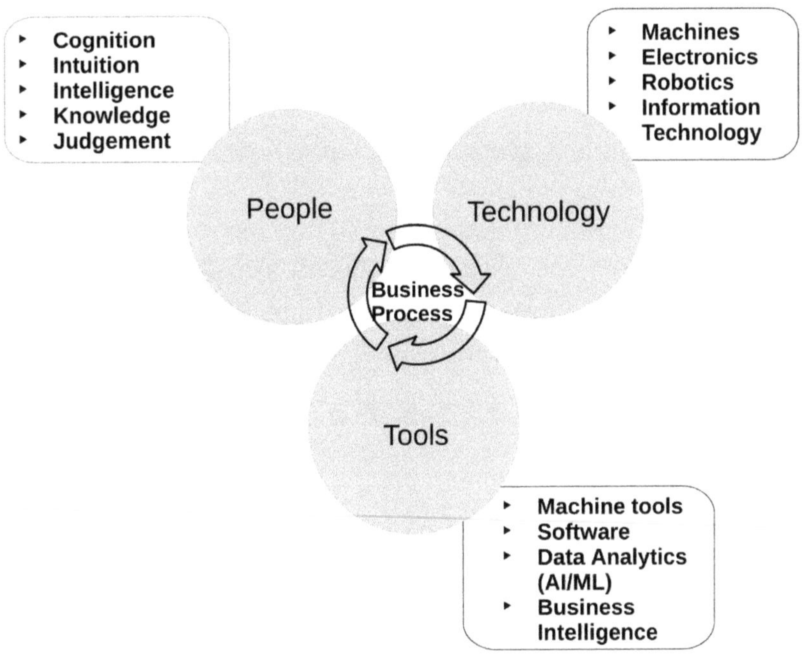

Figure 5: Business process orchestration with people, technology and tools

Augmentation or substitution of human tasks

It is clear from the annual reports of organisations that operating expenses on human resources is significantly higher than the IT spend, thereby highlighting the human resource-centric nature of organisational work processes.

Human resources has been arguably the most important component in the organisational activities and tasks. They provide unique sources of all knowledge, intuitive intelligence, innovation, flexibility and the ability to solve problems in unexpected ways by adapting to new situations. Skilled employees deliver economic growth for the organisations they work for. However, psychologists have identified limitations in human memory and cognition that prevent them from achieving optimal performance of tasks on their own. Hence, employees benefit from using technologies and tools which enable them to accomplish tasks and activities more efficiently. These include:

- instant access to vast amounts of information storage spanning multiple subject areas and topics without having to store it in human memory
- SharePoint servers for extending knowledge sharing across the enterprise
- email servers for effectively communicating without limits of distance or time.

Moreover, since the cost of performing computations has fallen at phenomenal rate, it provides strong economic incentives for business decision-makers to substitute ever-cheaper computing power for human resources where appropriate.

Deciding on when to augment and substitute human tasks requires knowing:

- results of detailed analyses and determination of how tasks in business activities and processes are organised
- how workers perform their job functions
- what skills they require
- how technology, tools and other resources can enhance expected outcome in business functions.

It includes measurements of improved productivity, efficiencies, effectiveness and their associated cost and benefits. It requires thinking about what magnitude of the business function's effectiveness depends on the right mix of tasks and resources involved in jobs, and how human tasks can often complement different technologies and other resources, rather than blindly assuming substitution.

Figure 6 and Figure 7 illustrate the above points, followed by examples from industrial manufacturing and customer service management. Figure 6 provides a high-level resource mix in varying proportions.

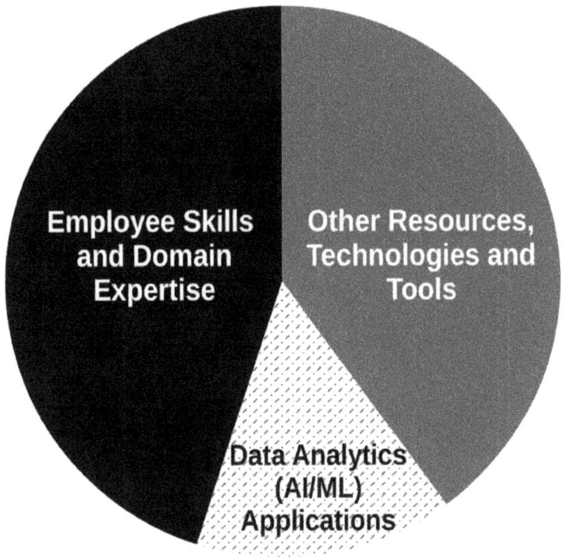

Figure 6: Effectiveness of business function with resources mix in work activities

Figure 7 contains next level detail of the resource mix proportions comprising employee skills and domain expertise, data analytics applications, and other technologies, tools and resources, which together jointly enable delivery of a business function's effectiveness in order to achieve the desired business outcome.

For instance, in situations involving routine and repetitive tasks accompanied by a stable environment, it provided comparative advantage when substituting workers with industrial robotics that led to automation, high productivity levels and growth in industrial manufacturing and production operation. Furthermore, when integrating sensor devices and computer vision in manufacturing equipment with robotics, they are found to provide higher levels of productivity and efficiencies by taking advantage of real-time data transmitted from the Internet of Things (IoT).

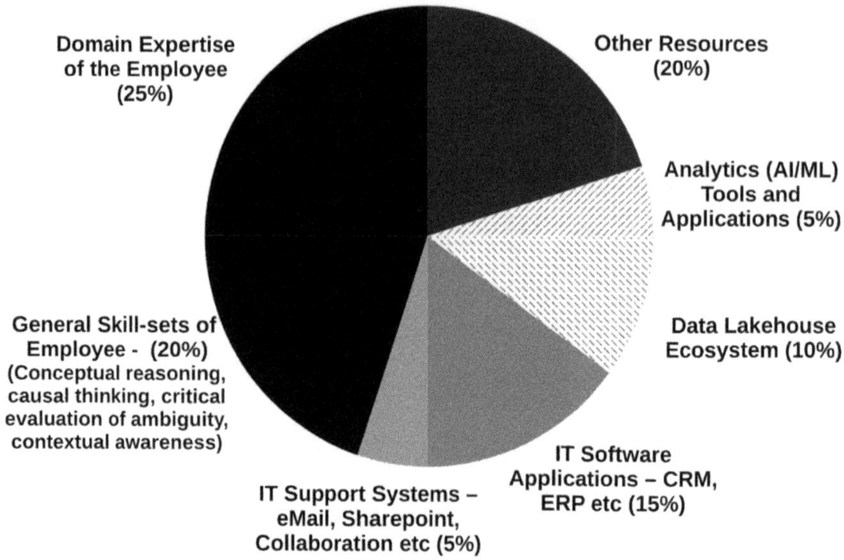

Figure 7: Distribution of resources mix in work activities (example from telecom and banking industries)

Emerging technologies, such as *Agentic AI* that combines artificial intelligence (AI), machine learning (ML), and large language models (LLMs) are expected to accelerate robotic process automation (RPA) by offering the opportunity to substitute low and medium skilled workers due to autonomous decision-making capabilities.

On the other hand, the unique strengths of humans, such as knowledge, cognition, intuition, originality and creativity, are essential for innovation and to flexibly respond to the unexpected by exercising contextual judgement and decision-making. Among other human skills, logical/analytical reasoning, problem-solving, emotional, social and communication skills play a significant role in organisational work processes.

Moreover, as the cost of processing becomes cheaper, humans will delegate machines to perform more frequent predictions and focus their attention on decision-making. The following example is a good illustration of the way technologies augments human tasks by taking advantage of data analytics technology while retaining business management, emotional, digital, social as well as cognitive and language skills to drive improved business outcome.

Consider a company that generates high volume and varieties of information. When a customer is talking to a contact centre agent, the customer's overall sentiment (happy/unhappy/likes/dislikes) or other information including customer's need, marketing opportunities, service issues and credit worthiness are generated by the CRM systems. It is not feasible to record and manually analyse this conversational and unstructured data spanning thousands or millions of customers. The company adopts data analytics that takes advantage of contact centre transcripts to assess customer sentiment and predicts the likelihood of the customer to escalate a service issue. This insight from data analytics helps the manager to proactively follow up on customer service and potentially avoids the risk of customer churn.

Prior to the adoption of data analytics, the contact centre manager conducted spot checks on a random selection of customers with service issues. The work required cognitive, social and communication skills and financial skills to identify potential churn-related revenue loss to effectively reach out to customers.

After the introduction of data analytics, the AI application analysed all the historical data to diagnose and predict priority list of customers on whom the manager should focus his attention. The job of the contact centre manager changed to emphasise greater customer interaction, retention and provide a coaching aid to improve the emotional and communication skill of contact centre agents.

Figure 8 illustrates how data analytics complement unique skills and capabilities of people in the organisation by supporting strategic planning and management, and business operations and management across the enterprise. Human cognition, intuition, conceptual reasoning, causal thinking and intelligence are applied when evaluating industry and market trends, which also helps in formulating the organisation's strategic direction and enables management in initiating improvement projects that are in alignment with business objectives. Similarly, unique reasoning skills and judgement under uncertainty/ambiguity enable business operation and management. The new insights and improved knowledge enabled by data analytics assists the operations management team to make decisions and take actions. Decisions and actions taken by the management team impacts business performance. Routine monitoring and tracking of business performance again feeds back into next iteration of strategic planning and business operation management.

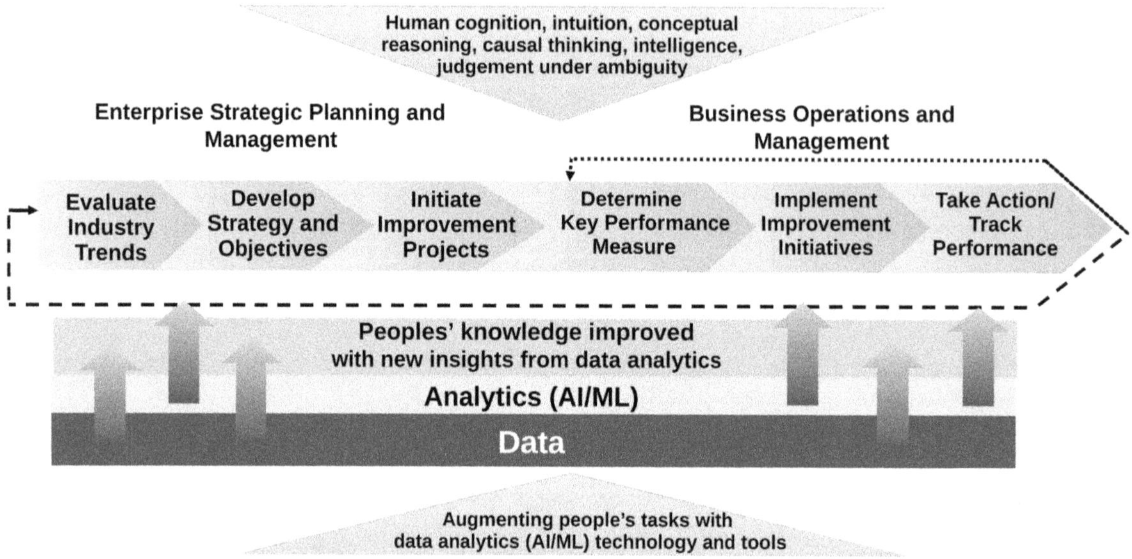

Figure 8: Augmenting peoples' tasks with data analytics (AI/ML) in the enterprise

The next few chapters will explore the resource mix employed in work processes across business operations and management. The objective is to determine how much of the business function's effectiveness depends on the extent and proportion of data analytics usage, among other resources, in delivering business outcome. This will form the basis of its value contributions to the business function and enterprise performance.

Part IV

How data analytics (AI/ML) drives business value

Part IV extends the topic from Part III by including details on the data analytics value chain with methods and processes for attributing data analytics value, among other resources in work processes. It includes illustrative examples from the author's experience related to healthcare and banking industries.

CHAPTER 6

Maximise value by aligning with business strategy

Overview: This chapter describes methods and processes for aligning data analytics (AI/ML) initiatives with business strategy. It also describes the data analytics value chain in detail with illustrative examples from the author's experience related to the healthcare and banking industries.

Data analytics (AI/ML) alignment with business strategy

The effectiveness of IT investments in general and data analytics in particular has been evaluated on one hand, using metrics, such as processing speed, efficiency and cost reduction. On the other hand, business units have been focusing on measuring business process improvement and enterprise performance by such measures as cash flow, revenue, margin and profitability.

The missing elements are methodology and process which show how business performance and results are being enabled by IT and data analytics technologies. Showing the linkage between data analytics technology, human capital and business performance is the first step in a cooperative process that involves consensus-building with the business stakeholders and users on metrics tracking.

Figure 9 provides an outline of the linkages between the different key components comprising human resources, business and data analytics. In particular, for data analytics investment to drive value, it needs to be linked to the vision, goals, strategy and long-term and short-term objectives of the organisation.

An effective way by which data analytics investments contribute to business performance is through its role in enabling business process changes in business improvement projects. Data analytics provides new insights that trigger a *light-bulb moment.* This can be spotting an emerging trend or an illuminating business strategy or discovering behavioural changes that are instrumental in enabling the management teams to make decisions and take actions. These include making process changes that sharpen the organisation's competitive edge. As a result, the value provided by data analytics can be measured using KPIs and by attributing its contribution to business and enterprise performance. In doing so, it needs to isolate the effects of other factors, such as people skills and other resources. Besides, when data analytics investments are aligned with business strategies and supporting operational objectives, the business teams will be motivated to participate in establishing baseline measures for continuous monitoring. Business teams may also identify new capabilities as a result.

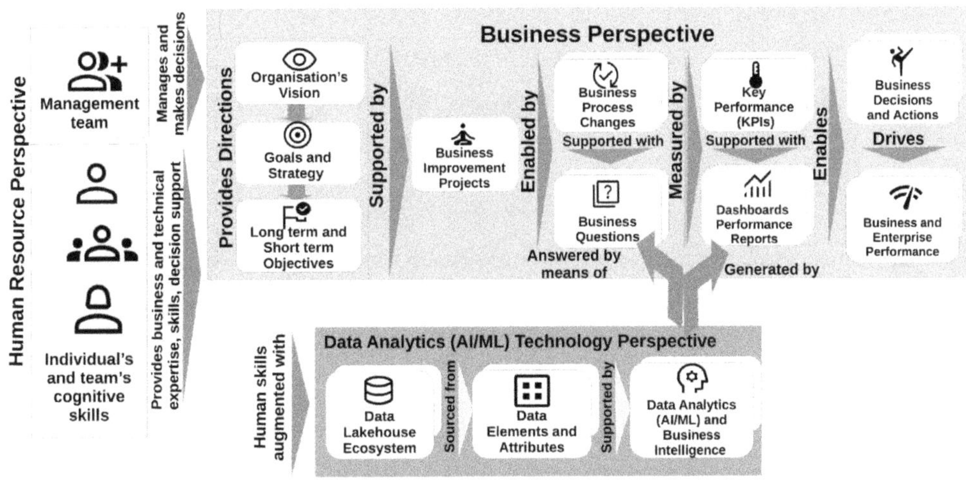

Figure 9: Alignment of data analytics (AI/ML) with human resources, business and technology

Data analytics (AI/ML) value linkage

Figure 10 further illustrates the linkage between business objectives, business questions, data requirements, machine learning and predictive modelling that provides new insights. The decisions and actions taken by business stakeholders from the insights provided by analytics are instrumental in business process engineering/changes that result in business performance improvements.

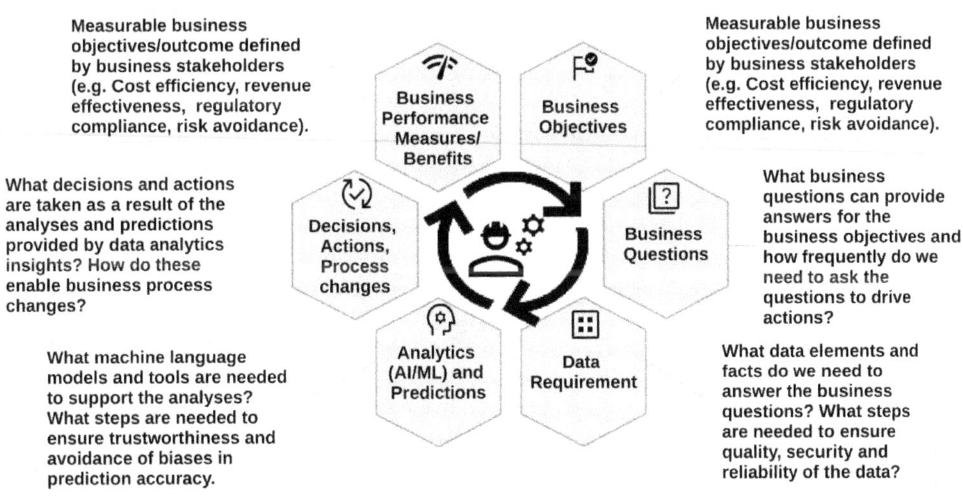

Figure 10: Data analytics (AI/ML) value linkage – How data analytics drives business value

Business objectives: Measurable business objectives/outcome, such as cost efficiency, revenue effectiveness, regulatory compliance, risk avoidance and quality of healthcare are defined by business stakeholders which provide an important basis on which foundational requirements for data and analytics can be established. If investments in data analytics are not aligned with the organisation's vision, goals, and business objectives, it can inhibit and even diminish business value by increasing the overhead costs on one hand and decreasing the contributions to business outcome on the other.

Business questions: Asking the right business questions can provide answers on the state of affairs of the business. It will also allow an organisation to track trends, predict and anticipate business performance, so that decision-makers can take appropriate actions for meeting their stated objectives and goals.

The type and sophistication of business questions increases at each level in the organisation and are directly proportional to the scope of roles, tasks and complexity of decisions involved as illustrated in Figure 11. Lower to middle levels of the organisation concentrate on operational performance, while at higher levels the executives focus on organisation strategy and enterprise performance.

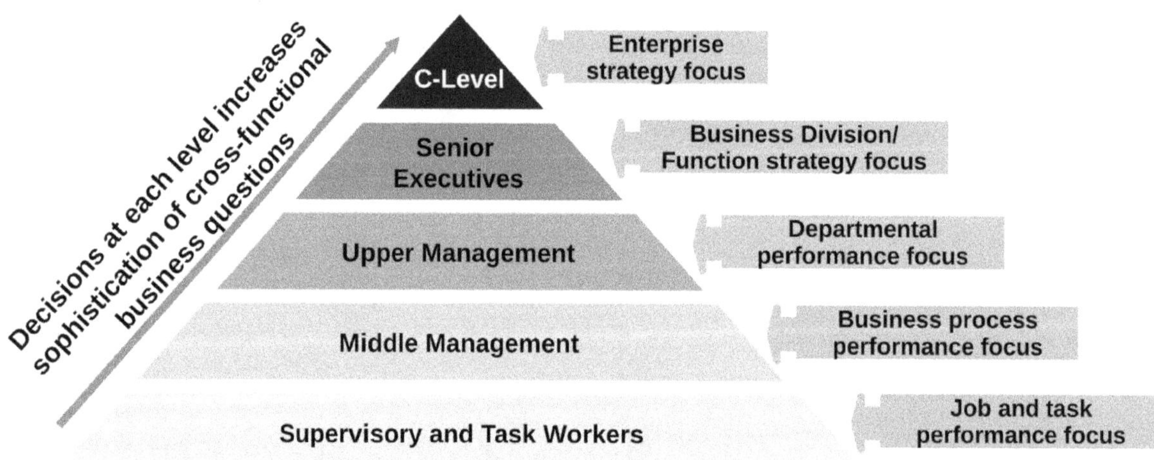

Figure 11: Complexity of decisions requires increased sophistication of cross-functional business questions

Workers and supervisors are generally concerned with decisions affecting performance of their jobs and tasks, such as,

'What are the contributing factors to smart meter failure?'

'Which of the smart meters has the best overall performance when balancing number of faults, equipment capability and equipment cost?'

They use data analytics to improve task performance and outcome of their jobs.

Roles in middle management periodically review business processes involving multiple activities and tasks. The degree of complexity in decision-making increases to inter-process efficiencies. For example, they deal with business questions such as,

'Which vendors are providing reliable equipment with high return on investment?'

'Which vendors should be introduced into the network?'

'What emissions, discharges, or other environmental impacts are trending with the potential future mix of new equipment and technologies being deployed in the network?'

Upper management is focused on departmental performance, while senior executives develop strategy at divisional and/or business functional levels and concentrate on cross-functional integration. The C-suites (the 'Chief' functions) develop enterprise strategy and focus on organisational level performance. Some of the complex decisions involve sophisticated business questions involving financial and business operations, such as,

'What FOREX risks are we exposed to and what hedging strategies should be used when acquiring a company with complementary products/services?'

'What policies are needed to reduce the risk of financial, reputation or other penalties associated with breaches in emissions regulations?'

Data requirements: Business questions provide the bridge between business objectives and the data required to answer the questions. The complexity, sophistication and the types of questions can change or increase the need for the data elements required to provide the answers. For instance, the question, *'What are the numbers and categories of prescriptions dispensed by pharmacy?'*, may need data elements containing the therapeutic class of the prescription, product detail, manufacturer information, and distribution channels.

Whereas, the question, *'What is the impact to drug utilisation when the company stops making the compound DGU951 for quality reasons?'*, will require more data elements, such as formulary levels used in the drug, inventory of the products involving the drug and compound, potential financial impact to clients and members.

As the number of data elements grows in the organisation, the need for effective data management and governance becomes more important. The way an organisation manages and governs its data impacts the business value.

Analytics (AI/ML) and predictions: To answer the business questions, a range of algorithms and tools can be applied in performing the analyses on the data. Descriptive analyses may require establishment of baselines, measure of variances and trends, while predictive analyses will require supervised or unsupervised machine learning algorithms to forecast the most likely future based on a sufficient reflection of the past.

Complex questions can be answered using natural language processing (NLP) and AI applications involving deep learning and Retrieval Augmented Generation (RAG) algorithms by leveraging Large Language Models (LLMs) and Generative AI (GenAI). These technologies are also useful in 3D modelling, simulation and augmentation using extended reality-based products and services.

Table 3 contains different categories and groupings of advanced analytical algorithms involving machine learning functions that are used by organisations for various business and industrial applications. Traditional supervised machine learning algorithms, such as classification and regression, are used to identify patterns for classifying and training existing data with labels to support predictions. This is typically used for applying them on new or unknown data. Besides, traditional unsupervised machine learning algorithms, such as cluster analyses (e.g. K-Means), creates groups with similar characteristics from existing data which is used to make predictions on new or unknown data.

However, AI applications perform tasks that require intelligence which is often learned rather than directly programmed. GenAI enables the creation and synthesis of new content in various text, images or other formats based on pre-trained LLMs that take advantage of advanced text analytics, NLP and deep learning using neural network algorithms. RAG algorithms enhance pre-trained LLMs by integrating relevant external data sources, such as an organisational specific data, to create context which helps to provide relevance for improving the accuracy of decision-making by leveraging conversational AI applications.

Decisions, actions and process changes: New insights provided by analytics applied on data elements increases the knowledge of business users and managers that enables them to make decisions and take actions, such as changing the course of negative trends based on new context. Using their expert knowledge and skills, business users and managers may simulate model results based on likely scenarios of different courses of action. Actions can trigger process changes in the organisation that align with business goals and objectives in order to drive the required business outcome.

Table 3: Various categories and grouping of advanced analytical algorithms

Anomaly Detection	Graph Analysis
Association Analysis	Journey Path Analyses
Attribution	Model Performance
Business Rules	Model Tuning
Classification	Neural Network
Classification and Regression (CART)	Optimisation
Cluster Analysis	Regression
Data Transformation	Statistical Analysis
Feature Reduction	Text Analytics
Fuzzy Match	Time Series
Geo-spatial Analysis	Visualisation

Agentic AI is an emerging class of software applications that mimic human intelligence and autonomous decision-making in solving complex problems. These applications exploit contextual external data and Retrieval Augmented Generation (RAG) to provide relevance. This helps the Agentic AI to adapt based on new context and feedback for achieving specific outcomes that are decided by predetermined organisational policies.

Business performance measures and benefits: There are a range of measurable benefits that can be expected from data analytics, including operational savings and new business capabilities. These are enabled by advanced data analytics with resulting impact to the top and bottom lines of the financial books of the organisation. Examples of business-focused measures include reducing risk through improved modelling of predictions, increasing revenue through uplift on marketing campaigns, and hyper-personalised customer experience –all reducing cost and waste through effective targeting. Data analytics contributes to these outcomes through rapid performance that speeds up the time to answer business questions. It also increases agility to make better decisions, because it not only provides easier, quicker access to quality data at various granularity of detail but also integrates with front-office and back-office systems across enterprise business divisions/functions.

Figure 12 provides an example from the healthcare industry on how data analytics initiatives can be linked and aligned to the visions, goals and business objectives. In this illustration, there are three key goals,

- Beyond quality to value
- Beyond hospital to community
- Beyond healthcare to health

These all form part of an overall vision for this healthcare organisation. Each of these goals have multiple business objectives corresponding to it that are linked to the goal. In turn, there are multiple data analytics initiatives to support these business objectives.

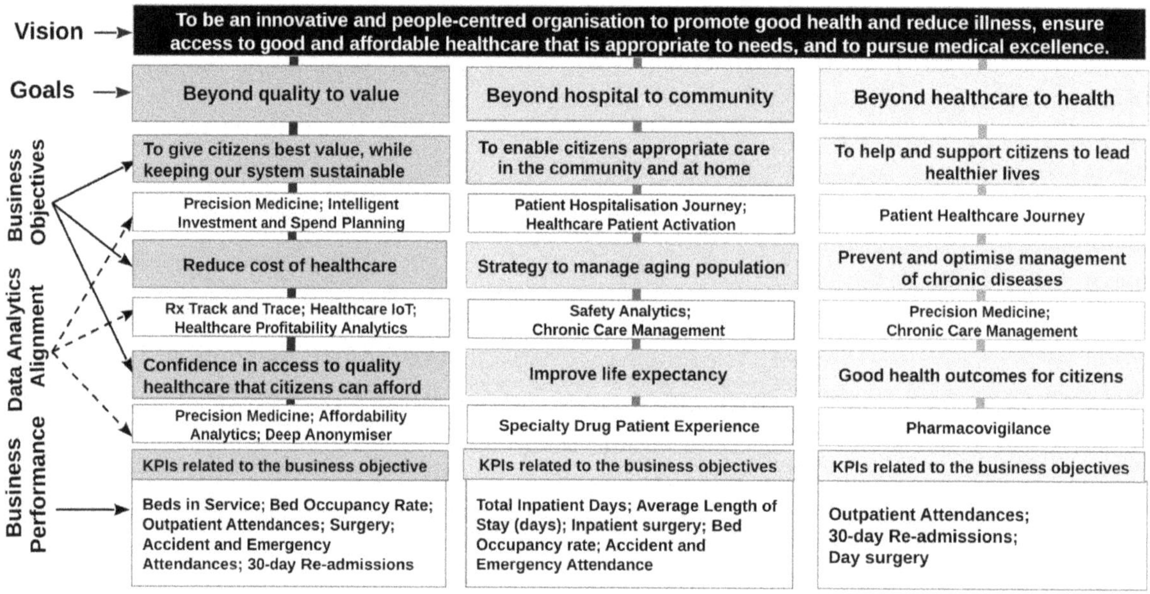

Figure 12: Data analytics (AI/ML) value chain – Healthcare industry example

For instance, Figure 13 contains examples of the data analytics initiatives 'Precision Medicine' and 'Chronic Care Management' that require behavioural analytics to increase the effectiveness of the business objective 'Prevent and optimise management of chronic diseases'. By analysing and matching the profiles of patients with genotypic, phenotypic, social settings and lifestyle of the population, data analytics tools help clinicians to develop proactive care plans to provide effective health services and care management. Data analytics initiatives also help in preventive care by predicting and avoiding patient re-admissions, thereby optimising costs while managing health outcomes effectively. The business value can be calculated by attributing the contributions of data analytics to the business outcome and key performance improvements. In doing so, it also isolates the contributions of people skills and other resources to the business outcome, in order to recognise the distinct value contributions of data analytics.

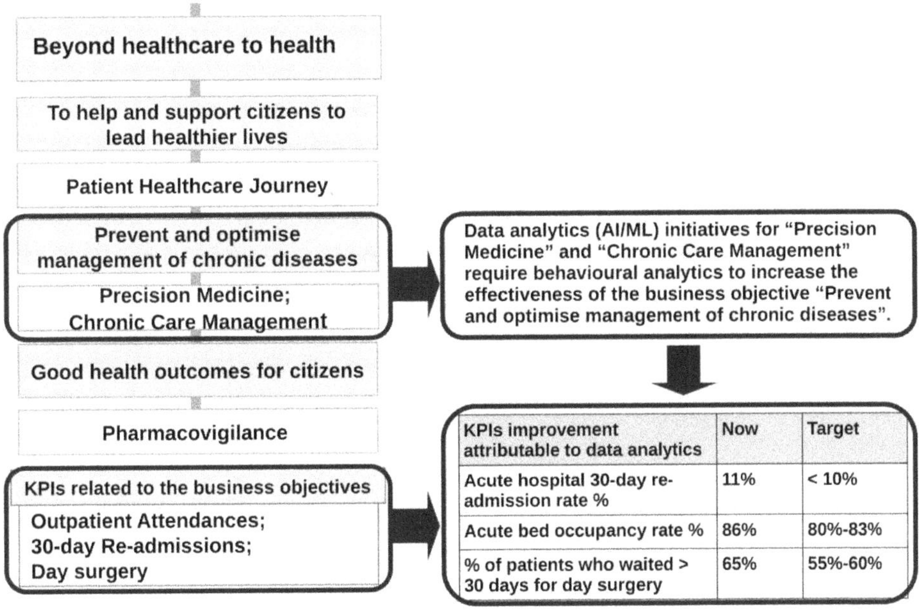

Figure 13: Data analytics (AI/ML) initiatives linkage to business objectives – Healthcare industry example

Similar to the above example from the healthcare industry, Figure 14 and Figure 15 provide examples from a tier-1 bank which illustrate the linkage between data analytics initiatives that are aligned with the organisation's visions, goals and business objectives.

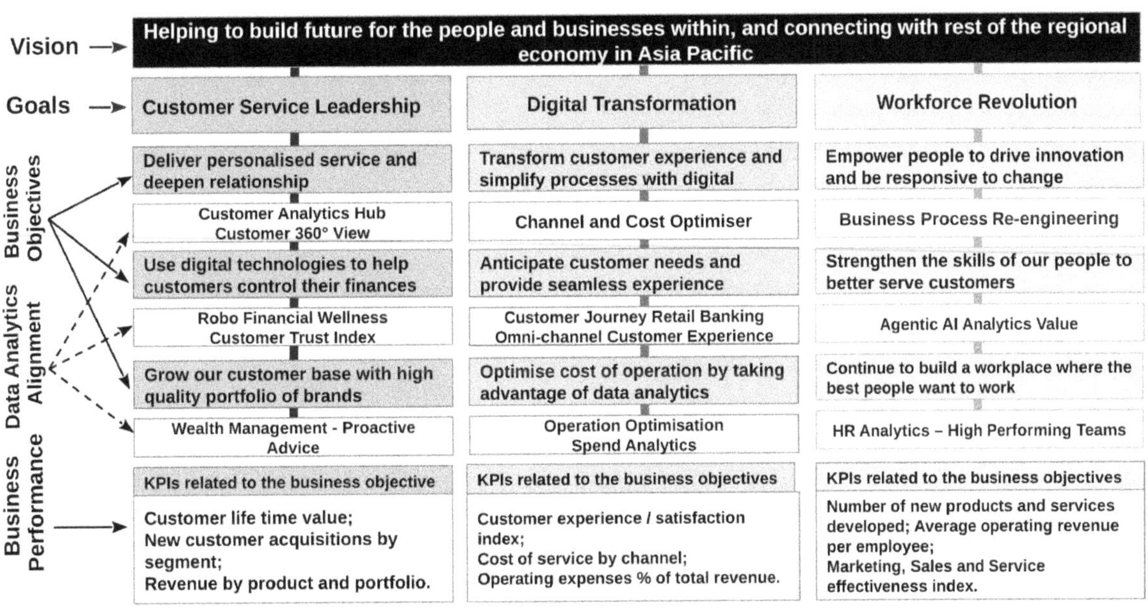

Figure 14: Data analytics (AI/ML) value chain – Banking industry example

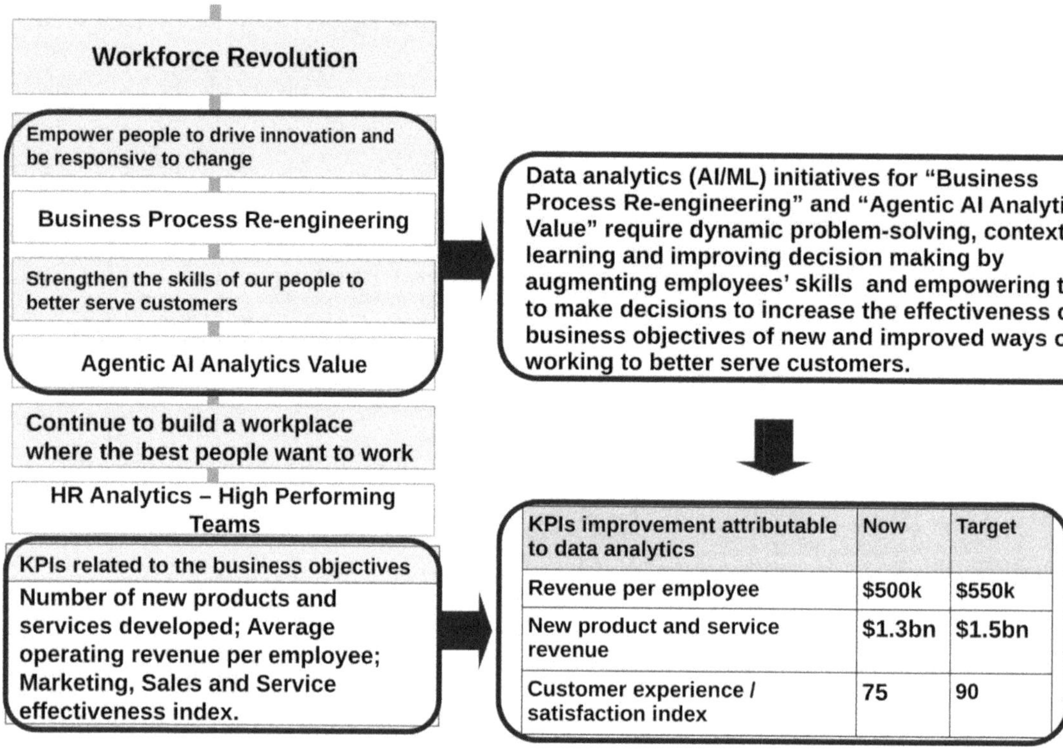

Figure 15: Data analytics (AI/ML) initiatives linkage to business objectives – Banking industry example

CHAPTER 7

Principles of data analytics value attribution

Overview: This chapter focuses on the principles of measuring and attributing data analytics value. It covers detailed analyses of how data analytics contributes to the business performance and business outcome of the business functions and sub-functions across the enterprise with illustrative examples for marketing and customer management functions.

Chapter 8 extends this to outline the organisational structure and business functions/ sub-functions that are common in many industries, along with detailed descriptions of the important role of data analytics in supporting a wide range capabilities to drive business performance across the enterprise.

Data analytics (AI/ML) value attribution

The linkage between business objectives, business questions, data requirements and machine learning and predictive analytics assists decisions/actions taken by management teams in delivering business performance improvements. As a result of this, the value provided by data analytics can be measured by attributing its contribution to business performance while isolating the effects from other factors, such as people skills and other resources.

Figure 16 outlines the value contributions of different resources involved in the business functions that drive business performance improvements and outcome in the organisation. This includes domain expertise and skills of employees, usage of data analytics in job functions, and employment of other resources and technologies in work processes, such as collaboration tools, knowledge sharing, and inter-functional communications infrastructure.

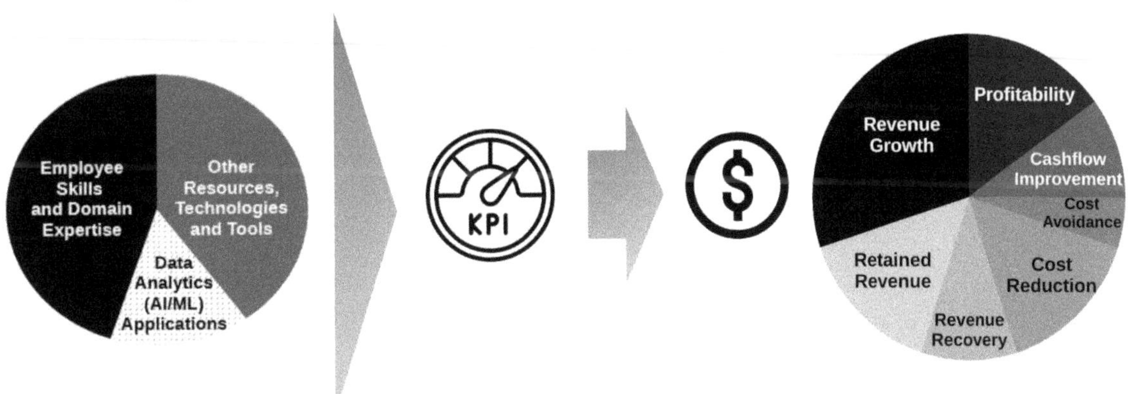

Figure 16: Value contributions of different resource mixes including data analytics

The value contributions of data analytics can be measured and attributed by asking the questions,

'What percentage of financial metrics/KPIs is the business functional area (i.e. sub-function) responsible for?'

'What amount of the business function's effectiveness depends on the data from the data lake ecosystem?'

'What other factors and resources contribute to your business function's overall effectiveness?' and

'What are the relative weights you place on the contributions of the insights provided by data analytics (AI/ML) towards the business function's effectiveness in management decision-making?'

By answering the above questions, we can calculate and attribute the value of data analytics by multiplying the business sub-functions financial metrics with relative weight of the contributions of data analytics involved in decision-making and key performance improvements as a result.

Figure 17 contains examples from Marketing and Customer Management business functions. These include Customer Behavioural Analytical Insights and Marketing Management, Product and Pricing Management, Customer Service/Experience Management, Sales, and Distribution and Channel Management. These business sub-functions are responsible for their respective KPIs of financial metrics (e.g. revenue growth, retained revenue, cash flows, profitability, revenue recovery, cost avoidance, cost reduction).

In commercial for-profit organisations, marketing and customer management business functions are predominantly revenue generating units of the enterprise with their KPIs largely focused on revenue (e.g. revenue growth, retained revenue, cash flows, profitability). Many of the back-office functions, such as Finance, Infrastructure and Assets, and Information Technology are cost centres with the main focus on cost management (e.g. cost avoidance, cost reduction, revenue recovery, cash flows, profitability). Almost all business functions are also responsible for cost-effective management of their functions and associated cost-focused KPIs.

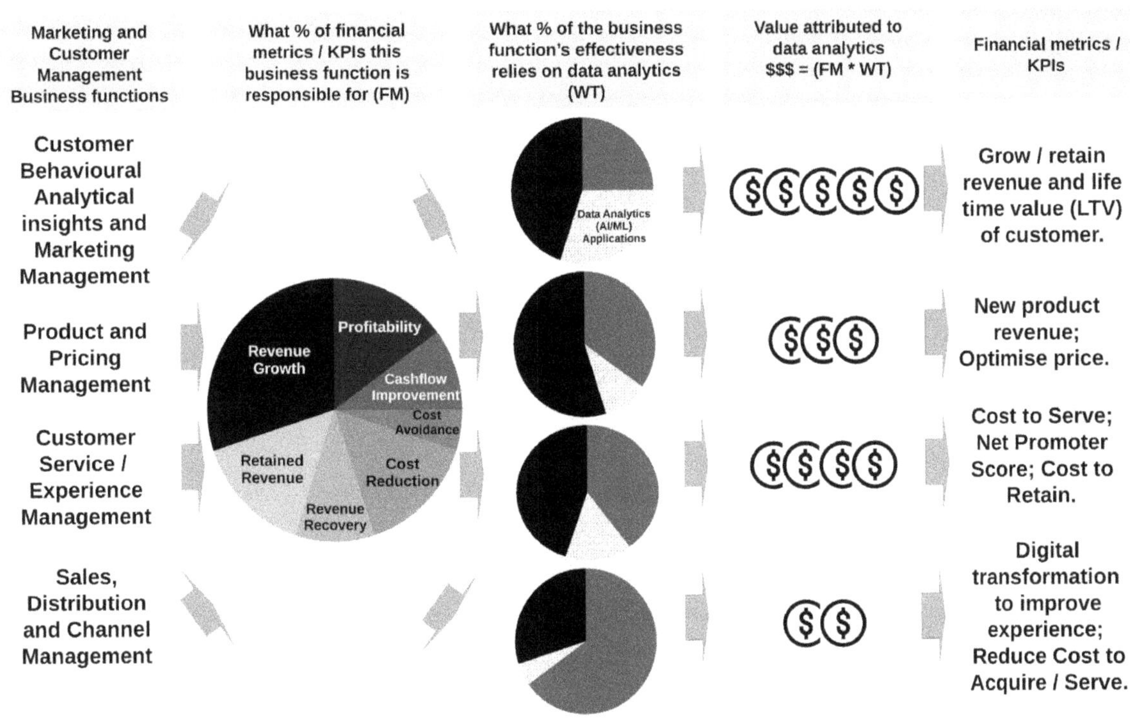

Figure 17: Data analytics (AI/ML) value contributions in business functions

Here are some examples of the KPIs for different sub-functions. The KPIs for the sub-function Customer Behavioural Analytical insights and Marketing Management generally revolve around growing revenue by means of cross-sell and up-sell activities in marketing campaigns. This business sub-function is also responsible for ensuring retained revenue by minimising customer churn and increase the lifetime value of customers by focusing on customer loyalty.

The Customer Service/Experience Management sub-function complements marketing functions described above by focusing on cost measures such as optimising cost to serve and cost to retain customers while also contributing to life-time value by improving customer satisfaction and net promoter score (NPS).

Product and Pricing Management provide support to the above-described business sub-functions by creating new products and optimising pricing of existing products by periodically monitoring market and competitive trends.

Similarly, Sales, Distribution and Channel Management sub-functions assist and sustain other customer facing business functions by providing a platform and managing the sales and distribution channels.

All these business sub-functions take advantage of data analytics in their work activities in different proportions. As a result, the business value contributions attributable to data analytics tend to vary as shown in Figure 17. The business value of data analytics will also vary and is subject to the capability level of the organisation in governing and managing data quality standards, privacy, security, centralised access, and talent in managing advanced analytics effectively and efficiently. Some industries tend to employ data analytics extensively in their business operations while others do so to a lesser extent. These will be covered in greater detail in the latter part of this book.

A simple principle for calculating business value of data analytics is by multiplying the business sub-function's financial metrics (FM) with relative weight (WT) of its contributions in decision-making and key performance improvements that results (i.e. $$$ = [FM * WT]). The relative weight can be assigned by involving the stakeholders and getting their agreement as part of value governance process which we will discuss in detail in Chapter 20, 'Data analytics value governance'. You can also take advantage of the benchmark weights across different business functions and sub-functions for banking and telecom industries – discussed in Chapter 18, 'Data analytics value estimation'.

Several examples of how data analytics business value is attributed are shown in Table 4 to Table 7.

Table 4 provides an example from customer service business sub-function.

Table 4: Data analytics (AI/ML) value attribution example – Customer Service/Experience Management

Business outcome: Increase in sales revenue for Value Added Service products and reduction of contact centre costs.
Business sub-function: Customer Service/Experience Management – Contact Centre
Data analytics value drivers: Agentic AI solution orchestrates dynamic workflows and brings together customer relationship management (CRM) functionality, and integrates customer's contextual historical data, and product data. Along with advanced analytics, it applies machine learning (ML) into a unified view and offer hyper-personalised customer experience management during inbound marketing. This has resulted in reduction of customer handling time by average 15 seconds per interaction while contributing directly to an increase in value added services product revenue of about 2.5% from cross-sell opportunities.
Business value of data analytics (AI/ML): Business value of $3.8m per annum with 20% weight attributable to data analytics (AI/ML) in achieving the operational objectives of $16m increase in value added service products sales and $600k in contact centre cost reduction.

Table 5 shows an example of customer behavioural analytics.

Table 5 Data analytics (AI/ML) value attribution example – Customer Behavioural insights and Marketing Management

Business outcome: Increase in retained revenue and reduction in customer churn
Business sub-function: Customer Behavioural Analytical Insights and Marketing Management
Data analytics value drivers: Micro segmentation, and precision targeting of customers was made possible by providing easy access to centralised customers' historical data, and social network activities data along with campaign management tools and Large Language Models (LLM) that provided capabilities for real-time engagement with customers. This enabled the campaign managers to anticipate customers' needs during interactions with them in a more natural, intuitive and meaningful ways to offer recommendations. One campaign delivered a 30% increase in bundled product as a substitute for separate products in the portfolio that customers previously bought and provided 15% savings for the customer. This resulted in increase in retained revenue and prevented customer churn.
Business value of data analytics (AI/ML): Business value of $1.05m per annum with 15% weight attributable to data analytics (AI/ML) in meeting operational objectives of 5% reduction in customer churn with retained revenue of $7m.

Table 6 illustrates an example from sales and distribution.

Table 6 Data analytics (AI/ML) value attribution example – Sales, Distribution and Channel

Business outcome: Reduction of cost to acquire customers and increase in revenue
Business sub-function: Sales, Distribution and Channel Management
Data analytics value drivers: Using the lifetime value of customers from the historical data in data lakehouse, combined with retail sales performance of best dealers/distributors, it was possible to optimise, and reduce the sales commissions paid to ineffective distributors. By using visual intelligence tools, and Geospatial analytics it was possible to visualise the location of high performing retail outlets to 'geo-fence' segments of high-quality prospects. This was instrumental in attracting potential customers to visit those stores which are effective in providing high quality service experience that influenced acquiring such customers.
Business value of data analytics (AI/ML): Business value $2m per annum, with weights of 10% and 15% respectively, attributable to data analytics (AI/ML) in achieving operational objectives of $15m consisting of $5m reduction of cost to acquire and $10m increase in revenue.

Table 7 contains example from back-office function dealing with revenue assurance in Finance.

Table 7 Data analytics (AI/ML) value attribution example – Finance – Revenue Assurance and Risk Control

Business outcome: Revenue recovery
Business sub-function: Finance – Revenue Assurance and Risk Control
Data analytics value drivers: The ability to access large amounts of integrated historical call detail records from logs in data lakehouse allowed the analysts to identify patterns of fraudsters and allowed recovery of lost International Interconnect revenues. In addition, it allowed assurance of future revenues while enabling suspension of the services for delinquent accounts. A combination of machine learning algorithms, also known as multi-genre analytics, was instrumental in identifying fraudulent patterns. This included fuzzy match, cluster analyses and anomaly detection.
Business value of data analytics (AI/ML): Business value of $1.5m per quarter with 35% weight attributable to data analytics (AI/ML) in International Interconnect revenue protection objectives of $18m per annum.

CHAPTER 8

Value chain in enterprise business functions

Overview: Following from earlier topics on the data analytics value chain and the principles behind the process for measuring and attributing data analytics (AI/ML) value, this chapter covers details of major business functions that are common in many industries and the important role played by data analytics to support business outcome.

Data analytics (AI/ML) applications across the enterprise

Data analytics provides the means to analyse facts and figures linked to the business performance of the business functions and the enterprise, and ultimately to the industry in which businesses operate. At the enterprise level, data analytics helps executives monitor KPIs, optimises resource allocation, uncovers bottlenecks and streamlines operations. For business functions, it provides new insights and supports decision-making which we will discuss in greater detail in the following section. At the industry level, it uncovers evolving consumer trends, market and competitive dynamics which provide an essential yardstick for developing innovative product and service strategies.

By applying data analytics in business processes, organisations achieve several quantifiable business benefits, such as more effective downstream processes, productivity increases that lead to resource and cost reductions, more precise targeting for better results, and development of innovative products and services.

Organisations are generally structured along the 'Internal value chain' business units that provide strategic planning and management along the vertical functions of strategic planning, assets and infrastructure management, and product/services development. The 'External value chain' front-line/customer facing business units operate and manage the business functions of distribution and channel management, sales and revenue management, and customer management. The front-line business units are supported by marketing function and supply chain management functions. The back-office functions are support services provided by human resources management, finance and performance management, information technology and data analytics, risk mitigation and management, legal and regulatory compliance, and corporate services management business units.

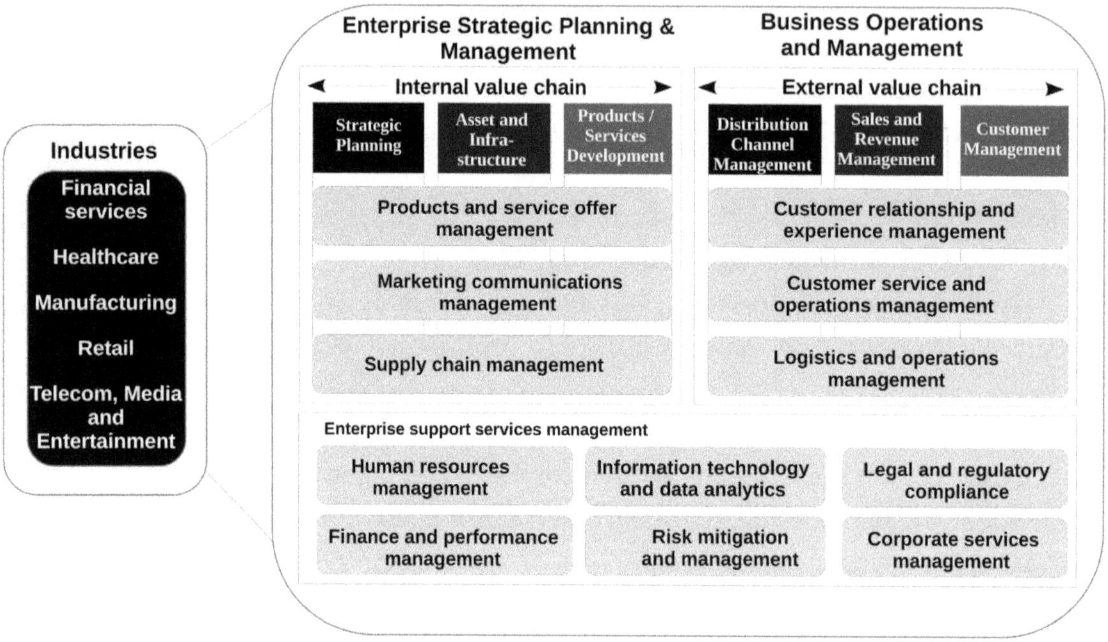

Figure 18: Common business functions/sub-functions in many industries (Adapted from © TM Forum, e-TOM Business Process Framework)

Figure 18 contains a high-level overview of common business functions and sub-functions that are common in many industries. The following section covers high level overview of how data analytics helps each of the business functions across the enterprise. Chapter 17, Chapter 18, Chapter 22, Chapter 23 and Chapter 24 contain detailed use cases and case studies from different business sub-functions covering industry examples. They describe some of the problems faced by the business sub-functions:

- how data analytics addresses the problem to achieve the expected business outcome
- what data is required
- what analytical techniques/algorithms are applied to generate insights/predictions and
- the business value contributions attributable to data analytics.

Strategic planning: In strategic planning, data analytics is utilised to examine the organisation's activities and resource capabilities. This includes capability assessment of workforce skills, technology, and finance for long-term direction of business operations consisting of infrastructure, product and services, distribution and channel, sales and revenue, and customer management. It is also used in analysing and aligning the organisation's activities to external environments, such as market and competitor dynamics, regulatory changes, and technological developments.

Assets and infrastructure management: Amidst demanding customer preferences, increasing expectations of service up-time, increased forces of competition, and the rising cost of capital, the need for investments in capital intensive industries is ever more important. Data analytics helps with the smooth running of the business by optimal management of physical assets and infrastructures from planning through procurement, operation, and disposal. Data generated by sensors and other components embedded in modern physical assets can be leveraged to perform advanced analytics to reduce costs of operation increasing service up time, while also optimising the use of human resources and increasing customer satisfaction through predictive maintenance. Besides maintenance, advanced analytical techniques assist in repair and operation, allowing businesses to sweat the assets and infrastructure that provide increased ROI. By integrating assets and infrastructure data with the enterprise and external data sources, data analytics aids decision-makers in a customer-focused approach for asset deployment and management.

Products and services development: Data analytics assists product development teams to recognise trends and drive strategic road maps for innovation, create new products and portfolios of services, source materials and resources, create catalogues, define and refine pricing.

Distribution and channel management: Distribution channels play a key a role in marketing strategy because they provide the means by which goods and services are funnelled from products and services development to reach customers. Applying data analytics for designing, controlling the process and optimising business operations in the distribution of goods and services across multichannels has become significantly more important. Data analytics helps to analyse and segment customers based on the product search and purchase behaviours of customers across different channels. These insights enable businesses to build strategies to drive increased sales and assist in the retention and quality of new customer acquisition through targeted marketing activities. A more detailed, deeper understanding of customers drives personalised communications to build customer loyalty and deliver improved return on marketing spend.

Sales and revenue management: Whilst the structural drivers of revenue for particular products and services are often well-established in sales and revenue management, there is often a wide range of factors which affect the achievement of revenue targets from sales. Organisations want to gain a better understanding of complex internal and external customer behavioural drivers which affect the revenue. High quality integrated data and advanced analytics help to accurately forecast, track and report revenue performance, identify revenue leakage and optimise product portfolio, product bundles and pricing to deliver revenue growth. Cross-selling and up-selling are an important tactic in revenue management that focuses on customers with high product holdings that not only generate more revenue but also help with loyalty and lifetime value. Data analytics helps marketers to identify the next best offer to be made to a customer based on their previous purchase history and current shopping/browsing behaviour to drive additional sales revenue. By leveraging advanced analytics techniques involving Geo-spatial, collaborative filtering and sentiment analysis, businesses can take proactive actions to maintain customer loyalty and protect sales revenues.

Customer management: Organisations are under increasing competitive pressure as customers seek better prices, quality and hyper-personalised service and these pressures impact their sales and margins. Businesses recognise that to achieve sustained growth, they must not only acquire customers but also understand customers' needs and wants to optimise their experience and develop long-term relationships. Customers expect organisations to know them, make interactions relevant, and provide an easy, consistent, seamless experience across touch points. In order to meet such expectations by customers, Customer Journey analytics has been employed in many organisations to provide insights on customer satisfaction levels that drive revenue growth and marketing effectiveness.

Products and service offer management: Data analytics facilitates the strategic process of analysing market performance to define and design products and services that customers want to buy, thus helping to promote and manage the sales and profitability. Data analytics also provides the ability to effectively manage distinguishing brand-building by leveraging the organisation's assets and infrastructure, and product and services portfolio. This attracts customers and maintains their loyalty, ensuring incoming revenue, profitability, and market share growth.

Marketing communication management: Over-communicating with customers can drive them to turn off and stop interacting with the organisation. Data analytics helps marketers to optimise contacts, prioritise communications and avoid contacting customers too frequently. Data analytics also enables segmentation of customers based on their attitudes and behaviours. This can enhance targeting capabilities and effectiveness of marketing messages, offers, incentives and communication by aligning to the needs and expectations of customers in those segments. Additionally, data analytics provides new insights by integrating customer profiles and feedback loops, allowing marketers to take advantage of digital channels to coordinate customer campaigns and interactions between online/offline, inbound/outbound marketing.

Supply chain management: Supply chain management is the planning and management of all activities involved in the supply, conversion and logistics activities to deliver the best possible service to customers. Data analytics assists with efficient supply chain management by enabling organisations to optimise inventory levels with high availability of products and services throughout the supply chain to deliver to customers in the most economical manner. It enables cost efficiency of logistics infrastructure and resources while delivering service excellence. A data-analytics-driven approach enables best practices in utilising a multi-functional approach that integrates sales and operations planning with cross-functional inputs from sales and marketing, finance and production.

Customer relationship and experience management: Organisations recognise the significance of acquiring and retaining customers in order to grow lifetime value and profitability. As a result, organisations prioritise lasting customer relationship and superior customer experience in the process of delivering products and services. Data analytics provides a comprehensive view of the customer journey, including multiple customer contacts across different channels. It enables organisations to improve business operational processes, ensure service reliability and reduce customer effort. Additionally, a predictive score and model of a customer's likelihood to perform an action (such as purchase a product, respond to a marketing message, or utilise an incentive offer) and enables personalisation to drive customer experience. The introduction of new products and services can impact product performance, quality and customer experience as a result of unanticipated consequences. Data analytics enables organisations to model the most likely scenarios and will enable more effective proactive planning and decision-making to mitigate the adverse impact.

Customer service and operations management: Customers often interact with organisations through multiple channels, even during a single conversation, such as a sales or service enquiry, a complaint or a claim that often results in multiple inefficient hand offs between agents and channels. The challenge is to understand the most common conversation types, then optimise the real time contact strategy to deliver a better customer service. Data analytics helps to advance customer conversations by exploiting real time interaction data across channels and agents. Data analytics plays a critical role in the rapid identification of root cause and provides proactive support in the prompt resolution of product/service quality issues to maintain customer satisfaction and loyalty. It also leads to reduced waiting and handling times when interacting with contact centres/digital channels.

Logistics and operations management: Retailers, distributors and manufacturers face changing demands which cause them to reconfigure logistics network in order to ensure that the right product is in the right place at the right time. Additionally, the increase of online, mobile, and social network channels put a strain on logistics costs and the ability to meet consumer performance expectation. Retailers want to understand how to configure their logistics network and optimise stocking policies to meet the increasing demands of omni-channel fulfilment, allowing them to optimise logistics costs, service and fulfilment lead time. Data analytics significantly contributes towards efficient and effective multi-channel, multi-path communication and logistics coordination of both product availability and transportation to ensure the correct distribution centre mix and transport for time-oriented customer delivery.

Finance and performance management: Apart from ensuring the financial stability of the organisation, there are core functions of finance where data analytics has been playing a pivotal role. Those functions are closing the books, profitability analysis, forecasting, margin analysis, budget and planning, timely and accurate financial reporting, accounts payable, accounts receivable, and treasury. Data analytics also enables finance managers to understand revenue, cost, margin and overall profitability at a granular level, which the General Ledger is unable to provide. For instance, data analytics provides revenue per customer, sales orders versus invoices per customer, levels of discounts provided per product/service per customer. Data analytics also assists with cross-functional analyses of costs, such as,

'What activities/events are the main cause of the distribution cost differences?'
'What is the distribution cost per product per customer?'
'Where do the cost variances originate?'

Human resources (HR) management: Human resources management is an area which often lags significantly behind other areas of the organisation in their use of data analytics. Sensitivities around access to and use of HR data means that it is often not analysed in conjunction with other operational and organisational data. Modern database security and tokenisation can support the required privacy and access controls, and the ability to apply sophisticated analytical techniques to HR and non-HR data can reveal many insights which can drive improved employee performance, recruitment and retention. With the emergence of AI applications that are potentially substituting humans in job tasks, the need for data analytics is significantly more important for HR managers.

Risk mitigation and management: Risk mitigation and management encompass the activities involved in limiting exposure and uncertainty to an organisation's ability to achieve their strategic objectives. The range of factors that can negatively impact an organisation are numerous, diverse, and complex requiring capabilities to mitigate their impact. Digital transformation has enabled increased visibility to many facets of the business but it has also created new vulnerabilities in employee and customer fraud. Data analytics helps many organisations to build profiles of fraudulent behaviour with transaction, account, inventory and product data to determine patterns and intervene before the event occurs. In-depth analyses predict potential risks and enable preventing actions to reduce their impact on the organisations.

In highly regulated industries, such as financial services, health care, energy, and telecommunication, data analytics plays a crucial role. In financial services industry, data analytics has been a core enabling technology in such areas as credit risk, liquidity risk, anti-fraud, and anti-money laundering.

Legal and regulatory compliance: In similar lines to risk mitigation and management, legal and regulatory compliance is a major function, especially in an organisation where data analytics has been critical in efficiently analysing petabyte volumes of data in compliance with regulatory reporting. The financial services industry complies with BASEL standards set by the Basel Committee on Bank Supervision (BCBS). Similarly, the telecommunication industry provides law enforcement agencies with communication network usage data and analyses.

Information technology (IT) and data analytics: Information technology management is a business unit in many organisations that provide the required technology infrastructure and systems as enterprise resource to all other business functions across the organisation. IT also provides data analytics capabilities, data science and analyses services to complement business analysts in business operations. Data analytics helps IT business units to manage the IT infrastructure and systems efficiently by enabling ease of access, service up time, governance and data quality to name a few.

Corporate services management: Corporate services management provides facilities to ensure business operations are efficient and effective. Similar to HR, this is an area that often lags significantly behind other areas of the organisation in terms of use of data analytics.

Part V

Evolution of ICT and data analytics

Part V reviews the evolution of the role of information and communication technologies (ICT) and data analytics in organisations along with the shifts in the roles of key stakeholders, their interests and their value perceptions, and associated value measurement approaches. We will also look at the evolution of the discipline and practice of value engineering.

CHAPTER 9

Evolution of data analytics (AI/ML)

Overview: This chapter reviews the evolution of the role of information and communication technologies (ICT) and data analytics. It also reviews the shifts in the roles of key stakeholders, their interests and value perceptions, and associated value measurement approaches. Much of the discussion in this chapter reflects the author's five decades of experience in ICT and data analytics. We will also look at the evolution of the discipline and practice of value engineering.

Evolution of the role of IT in organisations

At its core information technology (IT) consists of hardware, software and communication capabilities that enable organisations to gather raw data from paper records by mcans of manual data entry, collect data from digital sources, process the data, organise, store, distribute and transform it into usable information systems (IS) that support business functions and work processes across the organisation. Thus, IT together with IS have been instrumental in enhancing the knowledge of individuals and groups in the organisation by equipping them with better decision-making and organisational performance as a result. Improvements in productivity, increases in efficiency, reduction in costs, and ability to respond rapidly to competition have been some of tangible benefits of IT. However, without a concerted effort to manage IT and IS, **the intended benefits and value are unlikely to be realised**! It requires an understanding of the key stakeholders, their perceptions and interests on what they consider to be of value, inhibitors of value, the organisation's business objectives, and business processes.

Figure 19 illustrates the evolution of the role of ICT and IS evidenced during decades of experience in this discipline. During this period there have been dramatic shifts in centralisation and decentralisation of their architectures along with corresponding ownership structures in the organisation. The *data starved* mainframe era to the *data deluge* in the cloud computing era is also noticeable.

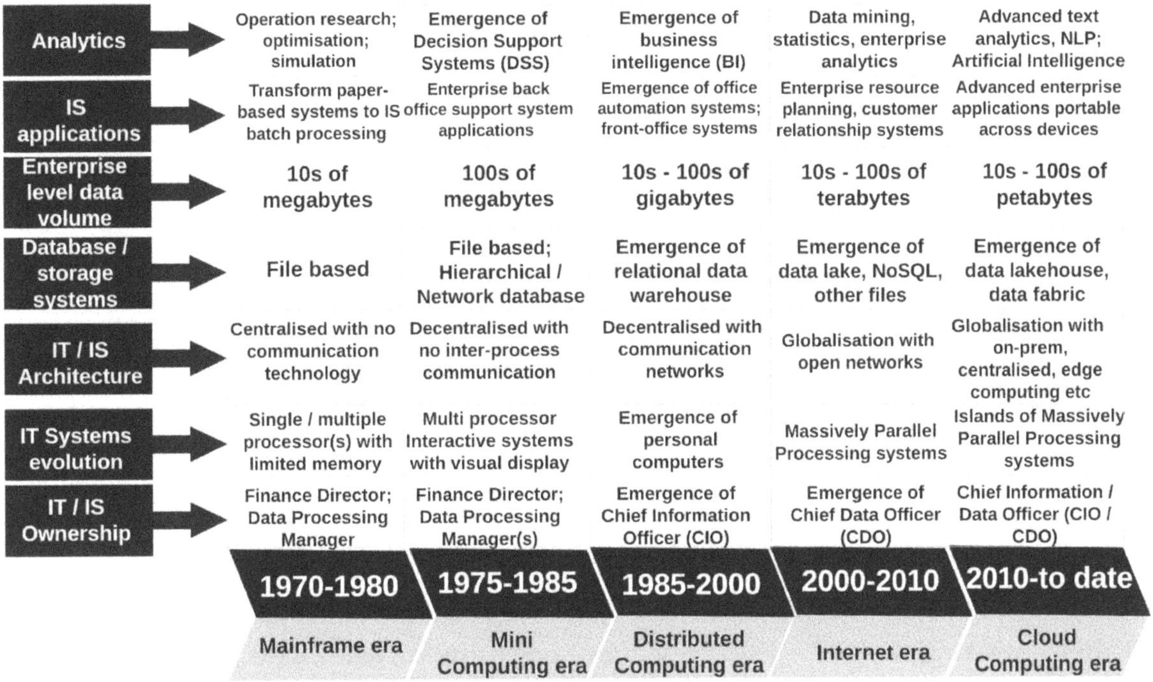

Figure 19: Evolution of the role of ICT and IS in organisations

Mainframe era (1970–1980): The mainframe era was characterised by centralised hardware that was massive in size, required climate-controlled rooms, and had very limited processing power and storage size. They were also very expensive to acquire and maintain. In fact, for many organisations, it simply provided an opportunity to convert manual paper-oriented processes into digital formats for capturing and processing financial reports. Commercial use of mainframe systems from International Computers Limited (ICL), International Business Machines (IBM) and Burroughs Corporation were common, with mainframe models such as ICL9001, ICL9004, IBM1401, IBM360, IBM370 and B1000. Some of these were single and dual processors with limited memory. For example, the ICL1901 had just 16K of memory. Online storage was on files limited to 2MB of removable disks that enabled sequential and random-access processing of data. The disk drive was like a huge washing machine! Mass storage of large files was on magnetic tapes with spool sizes of 600, 1200 or 2400 feet in length that provided sequential processing of data. Each tape drive was the size of a large fridge. Paper-based records, such as customer orders, invoices, payments, and shipping documents, were captured by data entry clerks on '*punch cards*' which were fed into optical character recognition (OCR) machines to convert into digital formats by means of Common Business Oriented Language (COBOL) or Formula Translation (Fortran) programs. Programs written in COBOL and Fortran were also used for subsequent processing for complex calculations, storing, and reporting. Data volumes were limited, largely due to the time it took to manually capture paper-based raw data as well as limited processing and storage capacity.

Business applications involved batch processing of warehouse/store transactions, sales journals, payroll, and financial ledgers used for management planning and reporting. Interestingly, analytical applications involved operations research (OR), optimisation and simulations, particularly for material planning in manufacturing and supply chain management. Ownership of IT and IS remained with the finance director with delegated authority for day-to-day management to a data processing (DP) manager. The DP manager maintained IT supplier relationships. The key objective has been to provide visibility and oversight of organisational performance to executives for their effective management, focused on efficiencies, productivity, and ROI for business operation.

Mini Computing era (1975–1985): The mini computing era had similarities to mainframes, but the hardware was mid-sized and affordable to large and medium-sized organisations. It also heralded an era of visual display units (VDUs) with keyboards called '*dumb terminals*' that were directly connected to the computers for capturing raw data without the need for '*punch cards*' which were common during the mainframe era. Mini computing was also an era where large enterprises could have many such mini computers in distributed locations for all data processing related geographically without the limitation of the centralised mainframe. This allowed for decentralised computing. However, there was no inter-process communication between distributed geographical locations due to lack of communication network infrastructures and/or huge communication costs. Data sharing between geographical locations was by means of offline transportation of spools of magnetic tapes. As could be expected, more applications were added to the portfolio of systems to support back-office functions and processes.

This was an era where hierarchical and network database emerged as viable alternative to sequential file-based storage and transactional processing. This was also a period when decision support systems (DSS) began to take roots in the organisation for preparing transactional and management reports. IT and IS ownership remained very similar to the mainframe era with the finance director in the business. DP managers were performing the everyday management. Computer operation was overseen by an operations manager. The key objectives and ROI metrics were very similar to the Mainframe era.

Distributed Computing era (1985–2000): The distributed computing era emerged as a defining moment in computing with the emergence of personal computers, office automation systems and data communication networks that enabled networks of distributed computers with inter-process communication and symmetric multi-processing (SMP). It also heralded an era of complexity in information processing that needed C-level executive management with the title of Chief Information Officer (CIO), which was first articulated by William Synnott and William Gruber in their 1981 book, *Information Resource Management: Opportunities and Strategies for the 1980s*.

This was also an era when relational database systems for transactional processing as well as for data analytics which emerged. Due to the shortcomings in transactional processing systems that were unable to cope with management reporting, data warehousing infrastructure, along with business intelligence (BI) applications, appeared as additions to the IT and IS ecosystems. This era also marked front-office support systems for marketing management and customer relationships, often supplemented with historical information from data warehouses. Data warehouses became central repositories with '*hub and spoke*' connections to transactional source systems on one side, and target systems and applications on the other side to serve the BI and reporting needs of the enterprise.

Data volumes in the organisation were still very small compared with present day, measured in 10s of gigabytes, or at best 100s of gigabytes. The Transaction Processing Performance Council (TPC) published its first 300 GB TPC-D benchmark results in 1998. It included companies like Informix and IBM. Towards the end of this era, TPC-D benchmark for 1TB (terabyte) was tested for Oracle database 8.0 on a Sun Starfire 10000 server to support high performance business intelligence applications.

In stark contrast to the earlier era, the distributed computing era divided the line between business function management and IT management, wherein the business units became internal customers of IT. This also resulted in changes to their respective objectives and key performance measures. The business units were focused on ROI, while ITs focus turned to total cost of ownership (TCO).

This era also began to widen the knowledge gap between business units' functional requirements, industry domain knowledge and the corresponding information technologies/ systems needed to meet and support the business requirements. Due to this widening gap between business users and IT departments, investments in IT and IS did not provide the needed returns and many of the investments have been unproductive and wasteful. The landscape became complex with compounding the problem of value measurements.

On the one hand, IT, IS and data analytics investments were considered as the necessary infrastructure and cost of doing business (TCO-focused overhead). On the other hand, they are expected to enhance productivity, efficiency and improve top line and bottom-line performance of the organisation (ROI-focused). Overhead costs are indirect business expenses that are not directly tied to the products and services that generate value. As a consequence, the TCO is under constant scrutiny for reduction of the overhead. As a result, the very investments in IT, IS and data analytics that are most needed to help organisations stay ahead in their market are at most risk from spiralling rounds of scrutiny for cost cuttings without considerations for its long-term value contributions.

Internet era (2000–2010): The internet era was a turning point in the evolution of ICT, distinguished by massively parallel processing (MPP) systems, emergence of fibre optic broadband networks and the internet protocol (IP) open network standards that enabled global connectivity with enormous, distributed processing power along with sharp decline in costs of information technology and communication networks.

This was also an era for the beginning of the open-source movement with the Apache Software Foundation led innovations, such as Hadoop and NoSQL, that led to the formation of data lakes with 100s of terabytes of storage capacity supported by MPP capabilities. The industry started using 3Vs (Volume, Variety and Velocity) to denote the frequency, rate, and a mix of structured and unstructured data (i.e. free format text, audio, images, video). This generated data that was previously unimaginable.

The internet era also provided an opportunity for organisations to expand the business applications to include enterprise resource planning, customer relationship management and regulatory compliance reporting. The data analytics ecosystem started to become complex with data warehouse, data marts, data lakes and file systems. The natural consequence of this evolution was also that most of the high volumes of data collected by digital means, such as the world wide web (WWW), was not usable due to redundancy. It needed better tools and solutions. Organisations started using data management tools, data mining, statistics and other analytics to effectively manage data to find useful new insights. Due to the complexity of the ecosystem as well as the recognition of the strategic value of data, it needed better governance and management oversight. The role of Chief Data Officer (CDO) began early in this era but adoption did not begin until around 2010.

Cloud Computing era (2010–to date): The cloud computing era is unprecedented in the evolution of ICT with its ease of access to data and information via ubiquitous availability of broadband and mobile networks and portable devices that have led to growth in user communities, social networks as well as corresponding increase in data volumes reaching petabytes in the organisation.

This also led to a greater divide in the IT and IS ecosystem leading to much more complex data architecture involving islands of MPP systems (i.e. data warehouses, data lakes, data lakehouse, data fabric, data mesh, etc.) in data centres on premises, in the central Cloud, and on the edge of the network. Together with expansions of business applications now easily accessible on any device from anywhere, this led to the need for managing data securely without compromise. Data analytics has now expanded to include advanced text analytics involving natural language processing (NLP). Agentic AI is starting to take root in organisations to substitute employees in front-office customer facing and some back-office roles. Along with this exponential growth, the problems of managing ICT have grown in complexity and there is expanded need for managing ICT.

Areas of specialisation have increased, ranging from hardware, networks, software, programming languages, data engineering, analytics, data science, Cloud, data centres, to centralised versus decentralised management. IT and IS management now involve the roles of CIO and CDO along with their individual responsibilities. The CIO is now focused on IT and IS infrastructure and operation while the CDO is responsible for driving business outcome, digital transformation, AI, and leveraging data as a strategic asset.

The challenges facing CIOs and CDOs include:

- optimum allocation of costs and resources
- balancing the sunk costs in legacy systems and technologies and new investments for modernising ever-evolving technology landscapes
- the need to integrate a large number of systems, technologies and data
- difficulties in sourcing specialist resources and skills
- widening gaps and disconnect between technology specialists' knowledge of the business/industry and the user community's needs and requirements, and
- lack of knowledge in the executive suite.

Evolution of value measurement approaches

Value engineering

Value engineering is a common approach used by many organisations for quantifying the business value of IT and data analytics investments. A Google search provides a quick summary of the concept and history of '*Value Engineering*' that appears to have its beginning at General Electric (GE) in the 1940s. It is an engineering approach to reduce the cost of physical product by recomposing the bill of materials (BoM) that makes up the product with less expensive components, but without sacrificing the functionality of the product. At its core, the method involves analyses of functions and features of the product, and the process of composing the components that go into the final product. A natural extension of this approach devolved into other disciplines in subsequent decades, such as total quality management (TQM), lean manufacturing, and Six Sigma, with the objectives of reducing defects, reducing waste, raising the quality of the end product, and improving process efficiencies and productivity.

Measuring the value by comparing ROI 'before' and 'after' improvements is a most common approach in productivity and efficiency improvement projects. But this does not work for data analytics investments because most common approaches assume a stable and predictable environment in which the causal relationship between initial problem and eventual outcome could be easily established. But not so in a data analytics ecosystem, which we will explore in detail in later chapters.

At each stage in the evolution of ICT, the need for managing ROI in ICT has grown in complexity, which many organisations find insurmountable. Besides, the changes in ownership of IT and IS at each stage in the evolution, along with the split responsibilities and perceptions of value between the business and IT stakeholders have widened, causing additional level of complexities. Moreover, the evolution of ICT has culminated in recent times with the rise of Agentic AI.

Part VI

Challenges in value measurement approaches

Despite the breadth of methods, approaches and techniques available for data analytics value measurements the challenges still remain. This part of the book explores these challenges, their root cause and impact, and formulates approaches for overcoming them.

CHAPTER 10

Challenges in data analytics (AI/ML) value measurement

Overview: This chapter reviews the challenges in data analytics value measurements along with their root cause and impact with the objective of formulating approaches for overcoming these.

Data analytics value foundation

Data can be considered as the foundational component of data analytics ecosystems on the basis of which information, knowledge, and value are created in the organisation. In contrast to data, information refers to a context created and communicated by the interpretation of data that has meaning to the user/recipient and can be understood by the user/recipient. This is where GenAI's and LLMs' shortcomings become apparent when integrating with an organisation's business processes and tasks, as they don't understand the domain-specific context of the industry, business functions and semantic meanings of the information to support business outcome. Model Context Protocol (MCP), Agent to Agent (A2A) protocol, Knowledge Graph and ontology are evolving techniques to overcome such limitations, but the problems still remain about semantic representation and considerations for synonyms, homonyms and standardisation of data models which we will discuss in Chapter 19.

Data must possess attributes of its quality that are fit for the purpose in solving business problems by enabling users/decision-makers to understand and make reliable judgements about business outcome. Figure 20 contains some of the key data quality attributes, such as accuracy, relevance and timeliness, on which business decisions can be reliably made in order to achieve expected business outcome.

Fit for purpose refers to the context in which data is used and its associated quality attributes. For example, data required and used for strategic planning tends to be broad in scope, has a future-oriented time span, and the frequency of data use is less. Also, summary data is generally acceptable rather than detailed data, and data does not have to be complete. Furthermore, data needed for operational decision-making needs to be up-to-date or current, can be historical, used more frequently, often narrow in scope, timeliness is critical and must be complete/comprehensive. For both strategic and operational use, data must be reliable/

trustworthy and must have measurable metrics that accurately represent an entity, process, event or place. Synthetic data generated for the purposes of training and testing should have many of the same quality attributes as in operational decision-making but its scope must be identified as such, and it must avoid data biases. However, for the synthetic data currency, timeliness, accuracy and completeness are less important.

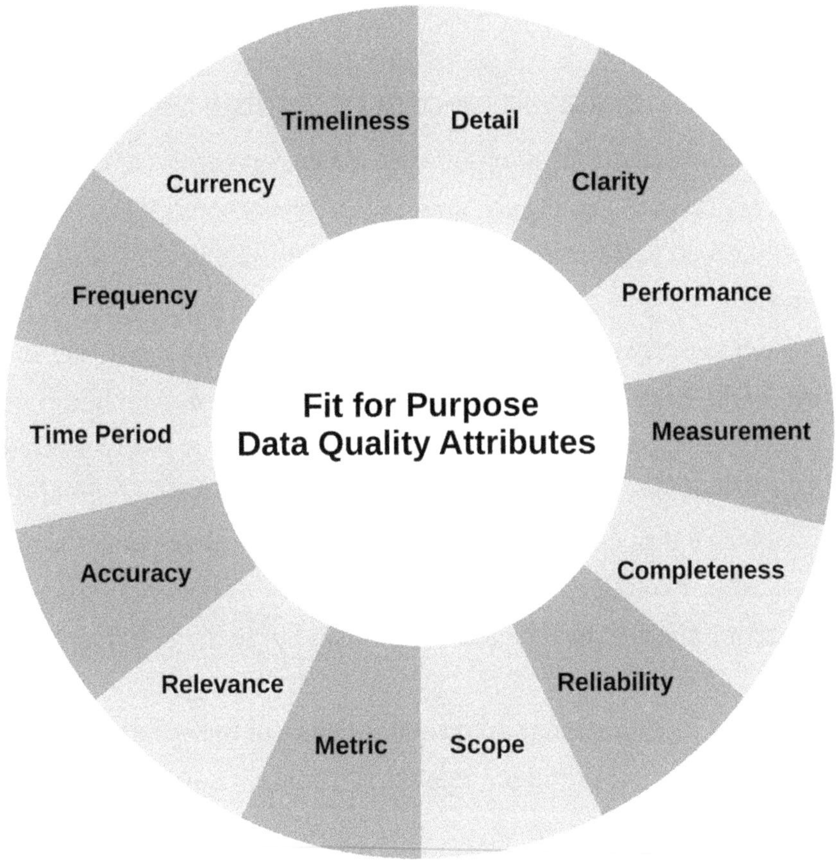

Figure 20: Fit for purpose data quality attributes

Departmental and application-oriented data stores inhibit value

Fitness for purpose is often impacted due to data redundancy and duplication of data. This results in data quality defects, such as uncertainty of the original data that leads to unreliability, inaccuracy, lack of clarity, obsolescence, and incompleteness. Departmental data marts and application-oriented data stores are often the root causes of such data redundancy and duplication, which directly impacts the business value by extending time to value and inhibiting competitive advantage. They also cause total cost increases as a result of architecture complexity, full cycle of development for each departmental data mart and/or application from data sourcing, data modelling, analytics development and integration with operational systems. Hence, the value created by data analytics depends largely on effective data management, data governance and value governance to ensure the quality of data is fit for purpose.

How departmental data marts inhibit business value

Asking the right business questions is important for ascertaining the true state of the business, so that workers, supervisors, managers and executives can track job tasks and business functional/organisational performance for taking appropriate actions for meeting its stated objectives and goals.

Because decision complexity increases at higher levels in the organisation than at lower levels, it requires sophistication of business questions being asked which in turn requires data from across departments, business functions and divisions to be easily accessible for timely decision-making. When data is stored at departmental or business functional levels, it leads to duplication of data from other departments or business functions in order to support business questions and decision-making. Duplication of data leads to data redundancy and increases efforts for integration, and thereby increases costs, delays time to market, and inhibits business value.

As shown in Figure 21, departmental data marts also limit the number of business questions that can be asked by limiting questions related to the data domain for the department or business function.

For example, without integrating any other data, Product data mart can only answer product related questions, such as,

'How much does shipping and other landing costs add to the unit cost of a product and how has this changed over time?'

'Which products are most profitable? Over the past few quarters, how has product profitability changed?'

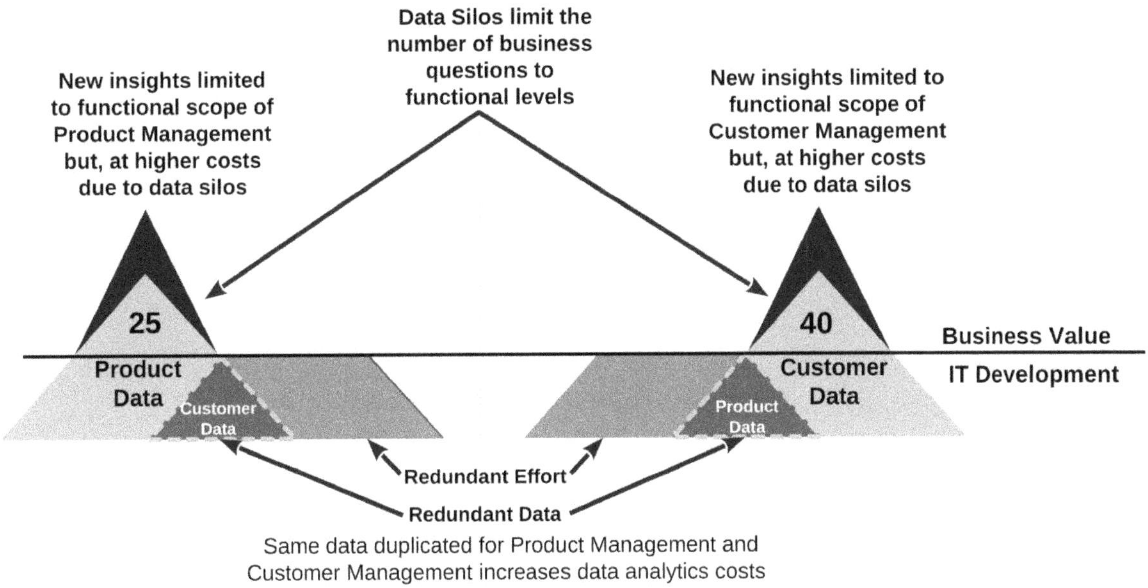

Figure 21: Data silos increase costs and inhibit business value (Adapted from © Teradata, 'Value of Integrated Data')

However, in order to find answers to business questions such as those below, it will require duplicating some of customer data alongside product data from the customer data mart and vice versa, because on its own the Product data mart would be unable to provide answers to such sophisticated questions. This not only increases IT development efforts, but it also increases total costs, delays time to market and reduces overall business value.

'*What is the cost of commissions paid for orders to convert a customer from one product to another?*'

'*What is product and customer profitability at customer and/or organisational level for value added service product line?*' and

'*Has converting a customer from an existing product to another product positioned the organisation to take a greater share of the customer's total spend (i.e. wallet share)?*'

On the other hand, when data from different departments and business functions are integrated into a common data repository such as data warehouse or data lakehouse or data fabric or data mesh, it increases the number of cross-departmental or cross-business functional questions at every level in the organisation and enables the organisations to ask more sophisticated business questions to make complex decisions as illustrated in Figure 22.

When integrating data from all relevant departments and business functions into a common repository, it avoids data silos, enables data sharing, and increases the number of sophisticated cross-functional business questions at each level. It also enables a large number of new insights for decision-making across business functions at all levels in the organisation with lower costs and increased business value.

Figure 22 provides a conceptual view of how data can be integrated into a common repository for data sharing. We can see that departmental data marts for product and customer on their own limits the number of business questions to 25 and 40 respectively. But, when product data is integrated with customer data into a common data store it multiplies the number of combinations of business questions to 125.

For instance, product data mart restricts product related questions, such as,

'*How does the spending for a product vary over time?*'

'*How much revenue was generated in the last month from various alternative funding products?*'

'*How accurate are forecasts at the product, location and store levels?*' and

'*What are the optimum levels for economic order quantities and safety stock given the historical demand?*'

Similarly, customer related business questions from a customer data mart may be restricted to

'*What is the price range on purchase orders for wholesale customers?*'

'*What are my receivable amounts in collections by customer?*'

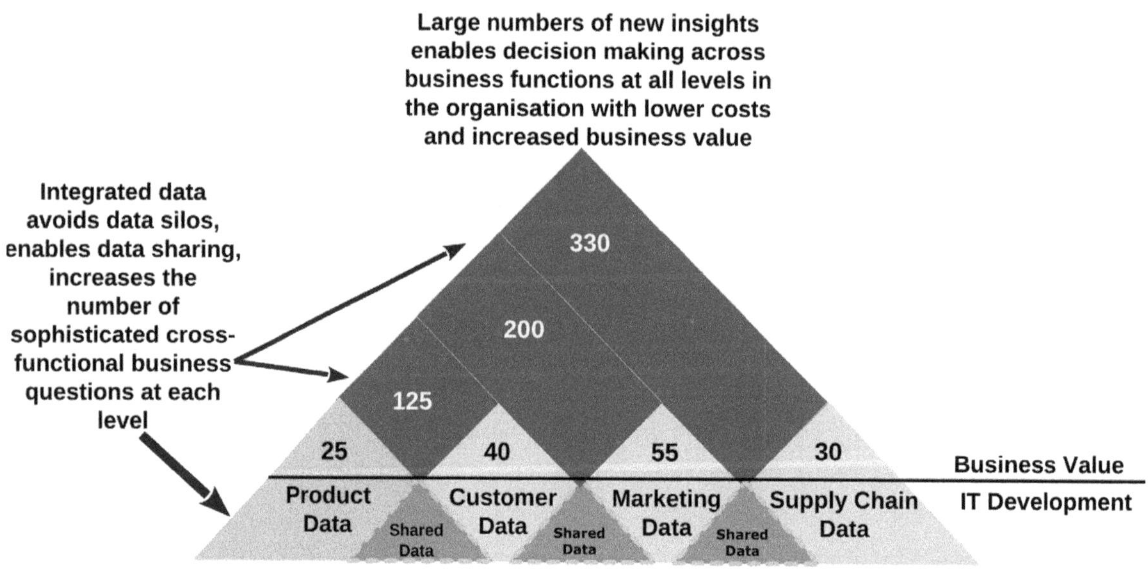

Figure 22: Integrated data reduces costs and increases business value (Conceptual view; adapted from © Teradata, 'Value of Integrated Data')

By contrast when integrating product and customer data, it broadens the range of combined questions, such as,

'What product is the customer most likely to buy next?'

'What is customer profitability by product at regional levels?'

'Has converting a customer from an existing product to another product positioned the organisation to take a greater share of the customer's total spend?'

Similarly, when data from product, customer, and marketing are integrated, the number of business questions increases to 200, and together with additional data from supply chain, finance, and so on, that are integrated into a common data lakehouse for instance, the level of sophistication and number of business questions climbs to 330.

Examples of sophisticated business questions that combine such multiple domains and subject areas include:

Product and supplier/vendor related questions

'Which margins are associated with product lines and suppliers?'

'What is my product cost from the vendor?'

'What are the costs of commissions paid for orders to convert a customer from one product to another?' and

'Given the replenishment and inventory status, what are current and projected excess inventory based on demand forecast, inventory levels, projected shipments, and service levels?'

Customer, product and marketing related questions

'What are the product trends by segment (volume, revenue, pricing and discounts)?'

'For complex multi-channel scenarios, what is the income range contribution by age group, by location?'

'What is the common customer journey for product acquisition by total number of products, by segment, and by product?'

'What are the top 5 demographics of customers that have various product mixes?' and

'Which customers are the best prospects for Investment Fund products?'

Customer, product and finance related questions

'For each account, what are the commissions paid compared to the contract write-down amount?'

'What is the correlation between commissions paid to sales managers and managing contract sign up most effectively?'

'For accounts in collections, what is the total outstanding balance?' and

'On the basis of delinquent payment history of instalment loans, how many, what percentage, and which customers average balances in credit card increased below 5%, between 5 and 10%, between 10 and 25%, between 25% and 50%, and greater than 50%?'

Customer, product, sales channel and marketing related questions

'What are the usage patterns of 5G services immediately after the purchase of a 5G handset and subscription (by segment, sales channel, mobile device model, subscription offer)?' and

'Which product and/or category will require larger than planned cash flow outlays to meet promotion needs or increased consumer demand due to expanded investments in 5G network over the next 15, 30, 60 days?'

In summary, by enabling sharing of data, data duplication and data redundancy are eliminated or reduced, which thereby reduces costs of development, speeds up time to market and increases business value.

Figure 23 provides an alternative view of data integration that employs the data mesh/data fabric concept that integrates the different data domains by means of data products supported by data analytics infrastructure, such as data lakehouse along with application programming interfaces (API) that achieve similar objectives by reducing data duplication and data redundancy.

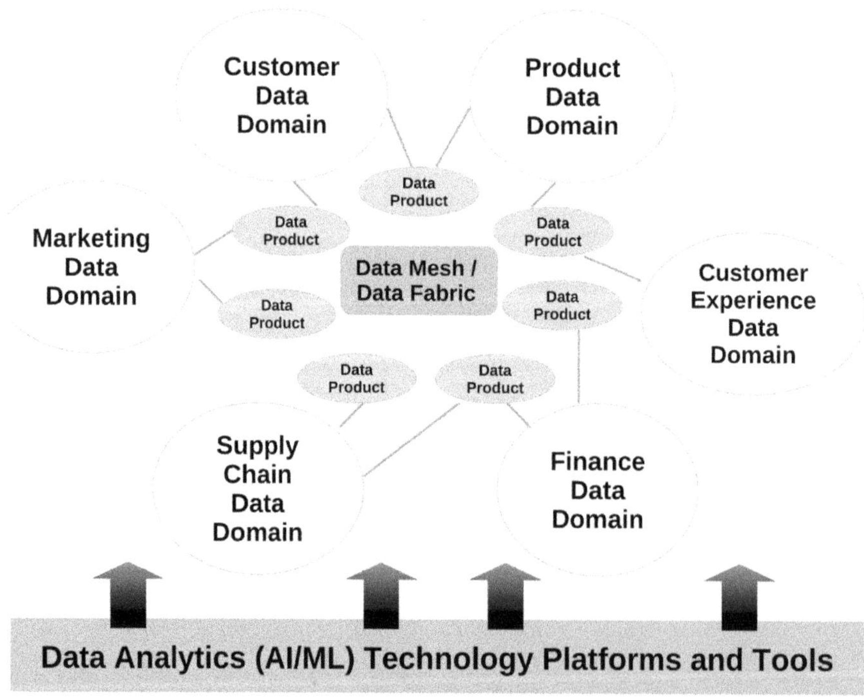

Figure 23: Integrating data in Data Mesh/Data Fabric by means of Data Products

How application-oriented data stores inhibit business value

How data redundancy and duplication causes bottlenecks and inhibits value has been superbly articulated by the head of analytics and modelling for consumer products at a tier-1 Australian financial services organisation:

'The data warehouse over here is "Application-oriented warehouse", meaning it grows via the execution of discrete applications which create duplicate data by ignoring what is already available, have their own objectives, constraints and data. The impact of this is that each application does not necessarily follow the same standards, which dilutes the "single source of truth" value proposition, and results in lack of reliability and trust on which decisions can be made. Additionally, it increases costs and diminishes value due to cleansing and reconciliation activities at our end'. **Consequently, many organisations accumulate technical debt.**

A senior manager for forecasting and reporting of a wealth management division said, *'IT can be cumbersome to deal with and are often distracted by "shiny new toys" without due regard for the business value they deliver or their impact on our existing architecture.'*

Data leverage saves costs and speeds up time to value

The way an organisation handles its data directly affects the preconceived and actual costs of a project. Some organisations are entrenched in the idea that data must be moved to applications, rather than designing applications to access a common data infrastructure. The result is data redundancy, higher aggregate cost of managing information assets, increased data security exposure, and greater data and information complexity.

Referring to Figure 24, in isolation these individual projects or applications may pass financial scrutiny, but in aggregate they may slowly destroy shareholder value. Likewise, when considered separately, a project or application may not pass financial scrutiny.

Figure 24: Without data leverage project efforts, costs and time increase

When leveraged against the efforts of other projects or applications, it may be viewed as feasible. Companies that have taken this tip on board quickly learn to leverage their existing data. As a result, they enjoy a dramatically different business case dynamic as shown in Figure 25 with lower total application costs and greater value realised faster. In this significant paradigm shift, not only do new applications incur marginal cost, rather than at their isolated total cost, but time to market can be significantly decreased.

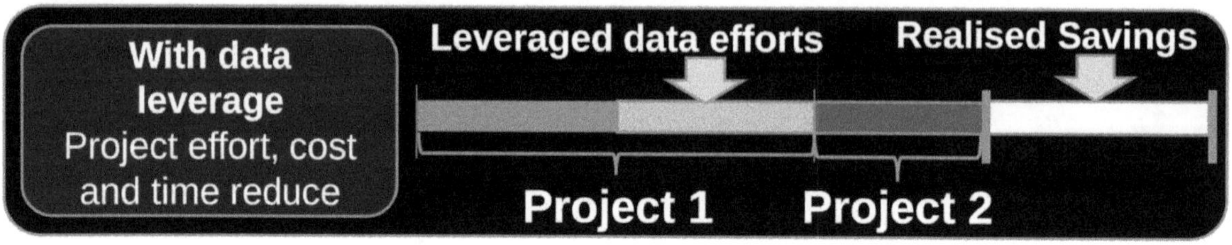

Figure 25: With data leverage project efforts, costs and time reduce

Data leverage with data reuse analyses

Data reuse and overlap analyses change how an organisation regards and values its information assets. While each project or application is considered incrementally and on its own merits in terms of its added financial value, they are together analysed in the context of data leverage and reuse for their cost and effort. This means that seemingly disparate projects or applications are rationalised in the context of data commonality and ability to leverage.

For example, in banking, telecom, retail and utilities industries, **Offer Optimisation** requires the product management team to demonstrate knowledge of its customers, demography, product pricing, buying behaviours etc in order to determine uptake of offers while maintaining profitability. Since much of this data is also used by other departments – including customer management, marketing and customer service teams – the data can be maintained in a common repository. This makes sense from both a business and IT perspective: lower cost of upkeep, plus increased consistency and accuracy.

If product management developed the application in isolation, it is unlikely that data would be shared. The project would therefore also have to take into account the cost of sourcing and integrating this data. Figure 26 provides an overview of how the product management team and other departments can leverage with data reuse analyses. It shows that data used for **Offer Optimisation** analytics is reused by 75% for Real-time Personalisation analytics, by 62% for Event Triggered Offers and Next Best Offer respectively, by 50% for High Value Customer Retention, and so on.

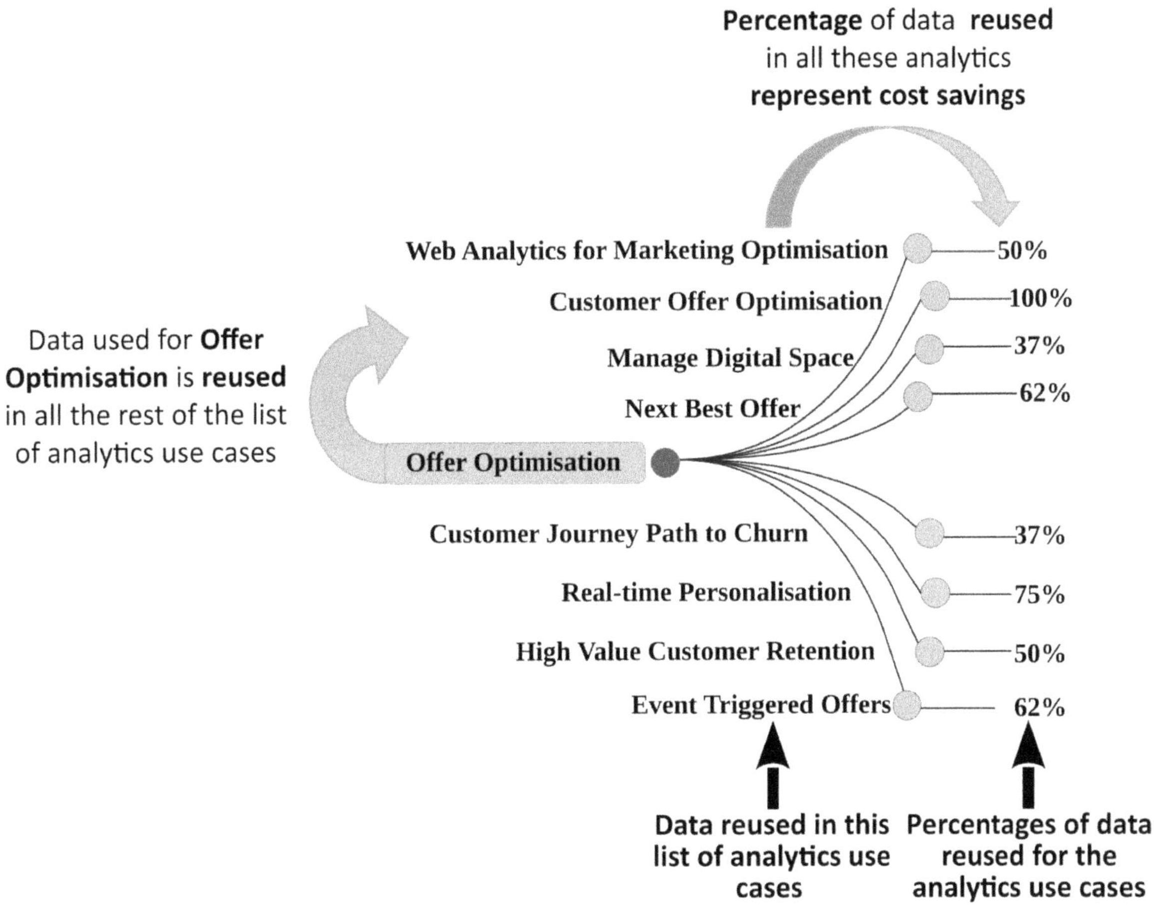

Figure 26: Data leverage with data reuse analyses

The cost and value metrics are shown in Figure 27. Without integrated data, each new application project (e.g. Offer Optimisation) must be considered and built from the ground up. First, new computing resources are sourced for each project or application, and database and network connections set up. Then the data model is established and new data extract, transform and load processes created. Analytics for the new data source is developed and tested. The data from source systems is loaded and integrated, and users are trained and granted access rights. In the end, all of this entails a significant effort that drives up the cost and lengthens the project's time to value. Using data that is not integrated forces the business case for this project to stand on its own.

Figure 27: Cost savings and time to value with data reuse from integrated data

With data leverage from reuse, the time, cost and effort associated with many of the setup tasks do not have to be incurred again. Instead, only the marginal cost of $200,000 is incurred by the other analytics that accrue an additional $700,000 of benefits while also shortening time to value by three months (i.e. six months instead of nine months). The result is a streamlined business that is better aligned with, and accountable to, shareholder desires to optimise investment.

Transactional systems vs. data analytics – differing approaches of valuation

Another challenge often encountered when measuring the value of data analytics investments is that many organisations apply the value engineering approach traditionally used for transactional systems, such as customer relationship management (CRM), sales and distribution (S&D), enterprise resource planning (ERP), financial accounting and performance (F&P) systems, which often limits the true long-term value of data analytics. These are operational systems that support day-to-day operation of the business either as front-office or back-office systems that interact with customers, channels, suppliers, partners, and materials/equipment on a transactional basis. Due to this, the focus of these systems is restricted to current and limited historical data. Another notable characteristic of the transactional systems is that they play a significant role in business process re-engineering and change management that are focused on business process improvement with specific key performance measures and indicators (KPIs) for productivity and efficiencies.

How these individual systems interact across the enterprise is best articulated by the value chain concept that shows the links in the value chain connecting suppliers, partners, channels and customers who are supported by their respective operational systems in the end-to-end chain. However, any process changes are effected in each system and are often measured with improvement metrics by comparing ROI 'before' and 'after' the changes. The business processes around transactional systems can be considered as stable and controlled environments, where the activities and tasks are well defined and streamlined. This helps in measuring the incremental changes and the corresponding business value contributions. By contrast, the data analytics ecosystem is in constant flux that would make it difficult to measure 'before' and 'after' changes to business value. Hence, it will require different approaches to measure data analytics value contributions. We will discuss this next in detail.

In comparison to transactional systems, data analytics systems provide decision support, business intelligence, predictive analytics, and reporting which take advantage of current and historical transactions generated from transactional systems such as CRM, S&D, ERP, F&P as well as many other non-transactional external systems and data sources, often known as BigData, to generate cross-functional and enterprise view for strategic planning. In addition, data analytics systems augment the CRM, S&D, ERP, F&P and marketing systems with new insights in support of decision-making when interacting with customers, suppliers, partners and materials/equipment. Data analytics in the enterprise is commonly designed as a centralised repository in an ecosystem, called a *Data Lakehouse,* that is a confluence of multiple systems supporting multiple users and tools with many months, if not, years of historical data and as well as future-oriented forecasts, projections and predictions.

Measuring the value contributions by comparing 'before' and 'after' improvements is a most common approach in productivity and efficiency improvement projects. But this does not work for data analytics investments because most common approaches assume a stable and predictable environment in which the causal relationship between initial problem and eventual outcome could be easily established. But not so in a data analytics ecosystem!

Data analytics ecosystems consisting of *Data Lakehouse(s)* is a confluence of hundreds of systems across most enterprise business divisions and functions (e.g. Customer Management, Marketing, Product Management, Business Operations, Enterprise Resource Planning), thousands of data sources, several petabytes of current and historical data and thousands of analytical models and algorithms that are constantly in flux which prevent value tracking and auditing. This problem is compounded by the agile software development approaches such as *DevOps, DataOps, ModelOps, MLOps* that instrument continual change to the data analytics ecosystem.

Challenges with tangible evidence for data analytics value

In many business functions and activities, measuring the data analytics value can be complex and difficult. This is largely due to the inability to link the value to tangible evidence. This can be in business functions and activities involving the application of data analytics for regulatory compliance, law enforcement undertaking, security/privacy protection, and other mandates wherein the investments in data analytics are considered as a compulsory cost of doing business with no measurable returns in its objective. This can also be in situations such as end-of-month closing of the accounting book where timeliness and completeness of closing the books earlier is important but, its ROI may not be evident. On the other hand, in the case of most investments in marketing, customer relationship management, revenue generating activities and transformation initiatives, it is relatively easy to attach value contributions of data analytics to the investments.

In Part VIII we will discuss in greater detail the methods and techniques for quantifying and attributing long-term business value of data analytics along with how to link value to investments. Also, we will look at methods for managing data reuse with illustrative examples from different industries across various business functions.

Part VII

Human vs automation with AI - An interdisciplinary framework

Part VII considers a different set of challenges arising in the organisation among administrators, analysts, administrative staff, domain experts, managers and executives in the context of unique capabilities of people, their cognitive biases, and limitations in judgements and decision-making.

Part VII further analyses these limitations and biases against similar uncertainties arising in data analytics in general, and the emerging applications of Generative AI (GenAI), and *Agentic AI* in particular as they relate to autonomous decision-making. Before discussing details on the limitations and challenges, we will also review how AI applications are likely to transform organisations by allowing the opportunity to re-engineer organisational processes to enhance productivity and efficiencies.

In Chapter 14* 'Techno-human dynamics in decision-making', Dr Jayanthi Raman, an academic clinical psychologist and researcher whose work centres on human neurocognitive biases, analyses the comparative strengths, and limitations of human reasoning and algorithmic systems. She outlines a set of recommendations and a strategic pathway for addressing the challenge of allocating investment between human and artificial intelligence in support of organisational judgement and decision-making. Her strategy emphasises maximising their respective value contributions and demonstrating how their combined impact strengthens business performance. She further examines contemporary understanding of artificial intelligence (AI), analysing how its learning architectures, predictive strengths, and cognitive constraints intersect with established psychological principles governing attention, emotion, and reasoning.

*Note: The original content for chapter 14 was authored by Dr Jayanthi Raman. She used AI-assisted tools (ChatGPT, OpenAI; GPT-5.3) to review, refine, error-check the language to improve clarity of the content she originally created, authored and approved.

CHAPTER 11

Re-engineering business processes with AI

Overview: In this chapter, we will review how AI applications are likely to transform organisations by allowing the opportunity to re-engineer organisational processes to enhance productivity and efficiencies. In doing so, we explain how AI will alter the resources mix in business functions and activities to perform tasks. While the capabilities of AI are unique, they should be considered as an evolution of information, communication and technologies (ICT) in the enterprise over several decades that have progressively enhanced productivity, efficiency, and created value.

Organisational transformation with AI

GenAI and *Agentic AI* are new generations of applications in data analytics (AI/ML) that are increasingly being deployed in organisations to automate decision-making traditionally performed by humans in the organisation. Simply automating activities and decision-making using GenAI and Agentic AI technologies will not be enough. These new technologies, together with existing domain specific software technologies in the enterprise, such as Customer Relationship Management (CRM) and Enterprise Resource Management (ERP), and predictive data analytics ecosystem will need to be integrated with organisational processes to deliver products, services and value by involving the business divisions and functions of the entire organisation.

For transforming organisational processes using GenAI and Agentic AI, the value chain model and business process re-engineering are effective approaches that we discussed in the context of digital transformation in Chapter 4. As part of business process re-engineering, we illustrated workflows depicting the sequence of steps in work processes along with tasks are performed in each step, and what resources are involved in executing the task, which will become very relevant when deploying the GenAI and Agentic AI technologies. However, as discussed in Chapter 7, determining which resources (e.g. labour, technology, tools) to invest that drive business value requires an evaluation of how tasks in business activities and processes are organised, how workers perform their job functions and what skills they require (e.g. technical skills, intuitive judgement, creativity, innovation). It also involves an assessment of what amount of the business functions' effectiveness depends on the unique capabilities, and what share of GenAI and Agentic AI technologies, among other technologies and resources, can enhance the expected outcome in business functions. Expected outcome of relevant resources deployed in business activities is measured by improved productivity, efficiencies, effectiveness and along with their associated cost and benefits.

Unlike other earlier generation technologies, GenAI and Agentic AI provide unique capabilities that could be leveraged during the transformation by considering work processes that could be fully automated with very little human intervention. This requires evaluation of the strengths and shortcomings of humans and AI technologies.

The section below, 'Evolution of technologies in the enterprise' explores how technologies contribute to productivity increase by enhancing the human tasks on the one hand, and substituting human tasks on the other when considering characteristics of work activities. Following this, we will discuss the unique capabilities and identify limitations of people in the section 'Human cognition'.

Then, in the section 'Mapping of task characteristics and skill requirements with human and AI/ML' we will explore in detail characteristics of routine, non-routine, manual and cognitive tasks in order to review how these characteristics match with capabilities of humans and data analytics. This will assist in formulating a plan for substitution and/or augmentation in business activities.

Following this, we will cover the skills, abilities and knowledge required in many of the job roles in the organisation, and identify unique cognitive, emotional skills and subject matter expertise of people that are essential for work processes. We will explore the limitations of AI technologies in the section 'Biases and limitations of humans and AI technologies'. This will lead us to discuss work processes and tasks that are ideal candidates for automation and autonomous decision-making as well as the associated investment needed in data analytics involving AI technologies.

Evolution of technologies in the enterprise

The widespread use of ICT in the 1970s followed by the ease of access to the internet technology in the 1990s has led to a dramatic increase in the value added per hour of work performed by people in the organisation. This is due to productivity growth as evidenced by 25% increase in Total Factor Productivity (TFP) and a 58% increase in worker productivity,[10] with a larger share of worker productivity expected from evolution of digital technologies.

Here are some of the ways technologies contribute to productivity and efficiencies. IT has been instrumental in processing of massive volumes of information and complex

10 João Amador and Cátia Silva (April 2023). 'The impact of ICT adoption on productivity: Evidence from Portuguese firm-level data', Working Papers, Banco de Portugal, no. wp202307.

mathematical calculations and has enabled augmentation of human cognition in sharing and exchanging of information. The internet enabled efficient and rapid searching of vast amounts of information, and further enhanced knowledge sharing. Electronics and digital technologies made it possible to sense, scan, process and transmit signals from RFID (radio-frequency identification) tags, process audio, image and voice signals for use in many applications that previously required manual tasks. In recent years, data analytics enabled predictions, optimisation, content generation, and a broad range of cognitive tasks that are capable of being used in systems and applications involving autonomy.

CHAPTER 12

Automation of decision-making

Overview: This chapter explores the aspects of human decision-making and the emerging trend of using AI technologies to automate decision-making in the organisation with their potential for substituting employees in the organisation. In the same context, we will discuss the unique cognitive capabilities of knowledge workers who have created tremendous value to the organisation through creativity, expert skills and innovation which the AI applications are attempting to mimic. The above discussions will form the basis of evaluating task characteristics, skill requirements of jobs, along with capabilities of humans and AI to determine which of these tasks could be substituted and/or augmented with AI.

Human cognition

Human work involves the cognitive processing of information from sensory signals to decide next steps in their intention or activities or to form a view of their environment which has been instrumental in producing results. On many occasions cognitive processing by humans is limited by a range of constraints, some of which have been overcome by technological evolution that has boosted human productivity along with its positive impact in the economy as evidenced by Gross Domestic Product (GDP).

The following examples provide a glimpse of how ICTs have complemented human tasks, and instances when technology can substitute for a human in processing information in work processes.

Example: e-Commerce

Since the advent of the internet, electronic commerce (i.e. e-commerce) has had phenomenal growth in the economy in online and mobile self-service buying and selling activities that were previously serviced by employees and sales agents in bricks-and-mortar retail stores, thereby substituting for human tasks.

Example: Automated traffic speed cameras

In Chapter 2 we reviewed examples of mobile speed cameras equipped with sensors that detect over-speeding vehicles by automatically scanning the speed of hundreds of vehicles to identify those that exceed the speed limit and/or cross an intersection that violates automated traffic signals. These tasks were previously performed by traffic police inspectors who have now been replaced by technologies.

Example: Health Services

Consider a mental-health clinician working with a client presenting symptoms of trauma. During the session, the clinician listens not only to the patient's description of events leading up to the symptoms of trauma, but also attends to tone, pacing, hesitations, emotional shifts, and non-verbal cues. They form an evolving mental model of the patient's condition, integrating clinical knowledge, prior experience, and tacit understanding of human behaviour. They draw on conceptual reasoning and pattern recognition to generate diagnostic hypotheses, which are then refined using the DSM-5 and their knowledge of comorbidity, risk factors, and contextual influences. Ultimately, they make a judgement about the most appropriate diagnostic formulation and treatment pathway.

While AI/ML tools cannot substitute for the clinician's expert thinking or the complex communication required in this interaction, they can augment both processes. Advanced speech-analysis systems can flag atypical linguistic patterns or vocal stress markers; vision-based tools can highlight micro-expressions or physiological signals; and case-based reasoning models can present probabilistic insights drawn from comparable client histories. These predictive systems can alert the clinician to potential diagnostic considerations or risk indicators, but the interpretation, contextualisation, and final judgement remain human responsibilities.

Mapping of task characteristics and skill requirements with human and AI/ML

Figure 28 maps task characteristics against human and data analytics (AI/ML) capabilities. It highlights that certain forms of expert thinking and complex communication remain uniquely human. These capabilities draw on originality, creativity, intuition, conceptual reasoning, and an ability to detect subtle or hidden attributes. They also involve interpreting ambiguity, establishing shared meaning, and exercising contextual judgement in situations where information is incomplete or rapidly evolving.

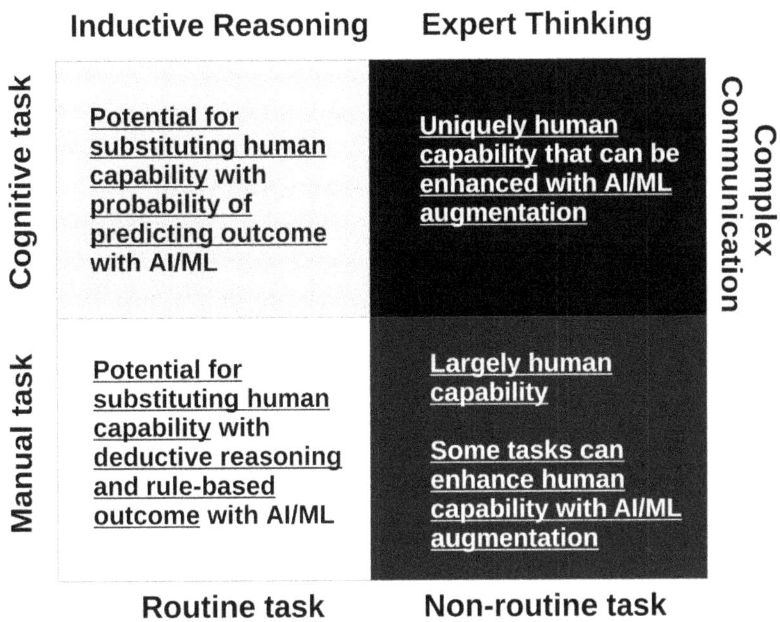

Figure 28: Task characteristics mapped with human and AI/ML capabilities [Adapted from © OECD (2010)]

The following provide descriptions of Figure 28, which have been adapted from OECD (2010).[11]

Tasks requiring **complex communication** consist of interacting with humans to acquire information, to explain it, or to persuade others of its implications for action. Expert thinking is a collection of specific solution methods that vary with the problem at hand. Tasks requiring expert thinking involve solving problems for which there are no rule-based solutions. One frequently used method is what cognitive scientists call case-based reasoning for constructing a new solution that begins with recognition of patterns of similarities and differences between the current problem and previously solved known problems.

11 OECD (2010), 'How technology changes demands for human skills', by Levy, Frank, 2010 Education Working Papers No. 45. OECD.

As Frank Levy[12] notes, *'Information is inherently ambiguous, and we give meaning to it by imposing a context. Without a shared context, there is no guarantee that the recipient of information will interpret it as the author intended.'*

Complex communication therefore depends on the uniquely human capacity to create a shared understanding, particularly in environments where information is abundant but context is fluid.

The examples covered earlier in this chapter illustrate the distinction between tasks that can be automated through prediction and rule-based reasoning, and those requiring expert thinking and complex communication, domains in which human cognition remains indispensable. While AI/ML systems can enhance decision quality by providing timely, data-driven insights, their outputs must be interpreted through human judgement to ensure diagnostic accuracy, ethical appropriateness, and contextual relevance. The interplay between statistical prediction models and expert human reasoning underscores the need for organisational strategies that harness the complementary strengths of humans and machines.

As Agrawal et al.[13] note, prediction refers to generating new information about an uncertain state of the world based on available data. Judgement, by contrast, relies on tacit, indescribable elements of human cognition (i.e. intuition, analogy, contextual reasoning, and the capacity to discern relevance amid ambiguity). AI/ML systems provide strong predictive capability, and humans provide the contextual judgement required for decisions. Effective decision-making arises from combining the strengths of both.

We will explore how statistical the prediction models work in the section 'Biases and limitations of humans and AI technologies', along with how defects/biases are introduced, and the steps needed to make adjustments to these defects/biases.

Routine cognitive tasks are mental tasks that are well described. Examples include evaluating credit risk of applicants for bank loans, evaluation of customer complaints to resolve faults, and investigating money laundering and identity fraud. Although these are routine tasks that can be executed with rule-based approaches for known problems, they often involve inductive reasoning and analyses to identify patterns of potential outcomes based on probability estimations. Estimations of probabilities can be done with historical data used to train models, such as regressions, neural networks, and other statistical models. These trained models can be

12 OECD (2010), 'How technology changes demands for human skills', by Levy, Frank, 2010 Education Working Papers No. 45. OECD.

13 Ajay K. Agrawal, Joshua S. Gans, Avi Goldfarb. (May 2018). 'Exploring the impact of artificial intelligence: Prediction versus judgement', National Bureau of Economic Research.

used for new cases to detect patterns and probabilities of outcome. These tasks are candidates for substitution by applications employing data analytics (AI/ML).

Routine manual tasks involve physical tasks that can be well-described using rules and deductive reasoning. Example includes inspecting road worthiness of vehicle safety. Other examples in agriculture include service robot application involving spraying insecticides, seeding, weeding and harvesting. Additionally, service robots[14] with an actuated mechanism programmable in two or more axes, moving within their environment with assistance from AI using gyroscopes, accelerometers and bump sensors data are increasingly used in households to perform useful tasks for humans, such as vacuum cleaning. Since these tasks can be defined with a set of precise, repetitive movements and rules they are candidates for robotic process automation (RPA) and data analytics.

Non-routine manual tasks are physical tasks that cannot be well described as a set of rules because they require visual recognition and fine muscle control that have proven extremely difficult to substitute with data analytics. Examples include hairdressers and other personal service providers and carers. However, some of these tasks could be augmented with robotic process automation (RPA) and data analytics such as assisted car driving.

Jobs and Task Characteristics

Jobs may demand varieties of skills to achieve business results. Table 8 and Table 9, which have been adapted from O*NET,[15] provide a list of skills, abilities and knowledge that are common in many of the job roles in business activities and work processes in the organisation. Table 8 provides a view of cognitive, emotional and social skills of people whereas, Table 9 contains general abilities, organisational and technical skills of people.

The implications of substitution of jobs by AI/ML application depends on the task content of jobs and the type of skills that people possess. Referring to Figure 28, routine task-intensive jobs are more likely to be substituted by AI/ML applications. On the other hand, tasks requiring non-routine, highly skilled expert thinking and involving complex communications require unique human cognitive capabilities and are less likely to be substituted by AI/ML application. However, the abstract tasks involving analyses and interpretation of information could benefit from augmentation of AI/ML applications. Nevertheless, it should be recognised that both humans and AI/ML have inherent limitations and biases which will need to be taken into account when evaluating substitution and/or augmentation of tasks.

14 https://ifr.org/

15 O*NET 29.3 Database by the U.S. Department of Labor, Employment and Training Administration (USDOL/ETA).

Table 8: Cognitive, emotional, social skills (Adapted from O*NET®)

Skills	Communication Skills	Customer Service Skills	Engineering and Technical Skills	Leadership Skills	Marketing Skills	Sales Skills
Active Learning	■					
Active Listening	■	■				■
Adaptability		■		■		■
Clear Verbal Communication	■					
Conflict Resolution		■		■		
Content Creation					■	
Critical Thinking				■		
Data Analysis					■	
Deductive Reasoning				■		
Emotional Intelligence				■		
Empathy		■				
Inductive Reasoning				■		
Judgement and Decision-Making				■		
Knowledge Of Industry Regulation and Standards			■			
Mathematical and Analytical Abilities			■			
Nonverbal Communication (Body Language, Tone, Etc.)	■					
Problem Solving		■				
Product Knowledge		■				
Relationship Building						■
Social Perceptiveness		■				■
Written Communication	■					

Table 9: General abilities, organisational and technical skills (Adapted from O*NET®)

Skills	Communication Skills	Engineering and Technical Skills	Leadership Skills	Marketing Skills	Sales Skills	Organisational Skills	Professional Skills	Scheduling Skills
Appointment Setting								■
Branding and Positioning				■				
Campaign Management				■				
Commitment							■	
Computer Programming		■						
Deadline Management								■
Delegation			■			■		
Feedback Delivery and Reception	■							
Goal Setting						■		
Integrity							■	
Lead Generation and Prospecting					■			
Market Research				■				
Motivation							■	
Productivity							■	
Proficiency In Software		■						
Project Lifecycle Management		■						
Project Planning						■		
Public Speaking/Presentation Skills	■							
Reliability							■	
Resource Scheduling								■
System Design and Implementation		■						
Task Prioritisation						■		
Team Building			■					
Time Management						■		

Biases and limitations of humans and AI technologies

Two aspects of biases are considered. This section discusses biases and limitations of AI/ML technologies in the context of statistical prediction modelling process. Chapter 14 will cover occurrences of bias from the psychological process of human cognition, memory, emotion, motivation, and context.

Statistical prediction modelling originated in the CRISP-DM (Cross-Industry Standard Process for Data Mining) framework by European Union's ESPRIT initiative. The CRISP-DM is still relevant for AI/ML, although the specific algorithms and techniques used in AI/ML are different (e.g. supervised and unsupervised machine learning algorithms; neural network deep learning algorithm etc). Also, modelling for conversational AI/ML in Agentic AI applications also requires dynamic iterations of goal setting, policy determination about autonomous decisioning, prompt engineering, model training/learning, adaptation within the organisation's workflow processes, and oversight for ethical governance and bias mitigation.

Figure 29 contains an adapted version of the CRISP-DM framework that illustrates the human in the loop (HITL) process of developing prediction models. Sometimes data analytics (AI/ML) prediction models produce unfair outcomes, known as bias. At various stages of the modelling process, biases[16] appear with unintended consequences. The following provide descriptions of the model development process along with the type of biases that can occur. Many of the biases are human memory, cognition and motivation induced which we will explore in Chapter 14.

According to Siddique, S., et al. '*Bias surfaces in different ways. Problems are frequently caused by choices made by people who develop or train machine learning algorithms. They might create algorithms that exhibit consciously or unconsciously biased thinking. Conversely, humans can introduce bias by using biased, erroneous, or incomplete datasets to train and/or validate machine learning algorithms*'.

Business Understanding: Any data analytics model development must be aligned with the business strategy and objectives. A clear definition of the problem statement along with expected outcome must be stated, which will form the basis for next steps in the process. This step in the process is dependent on industry, domain knowledge and expertise of people in the organisation. This includes participants from the business who will have industry and business functional expertise, and data scientists who will have data and technical knowledge as well

16 Siddique, S.; Haque, M.A.; George, R.; Gupta, K.D.; Gupta, D.; Faruk, M.J.H. (December 2023). 'Survey on Machine Learning Biases and Mitigation Techniques', *Digital* 2024, 4, 1–68.

as the ability to translate business understanding into common understanding of data and information representative of the business problem. However, when the data scientists gather business requirements from problem statements, their understanding can inevitably led to 'Cognitive bias' influenced by their past experiences and shortcomings in knowledge of the industry/business domain. 'Cognitive bias' can also be due to the challenge of relevancy in integrating much information.

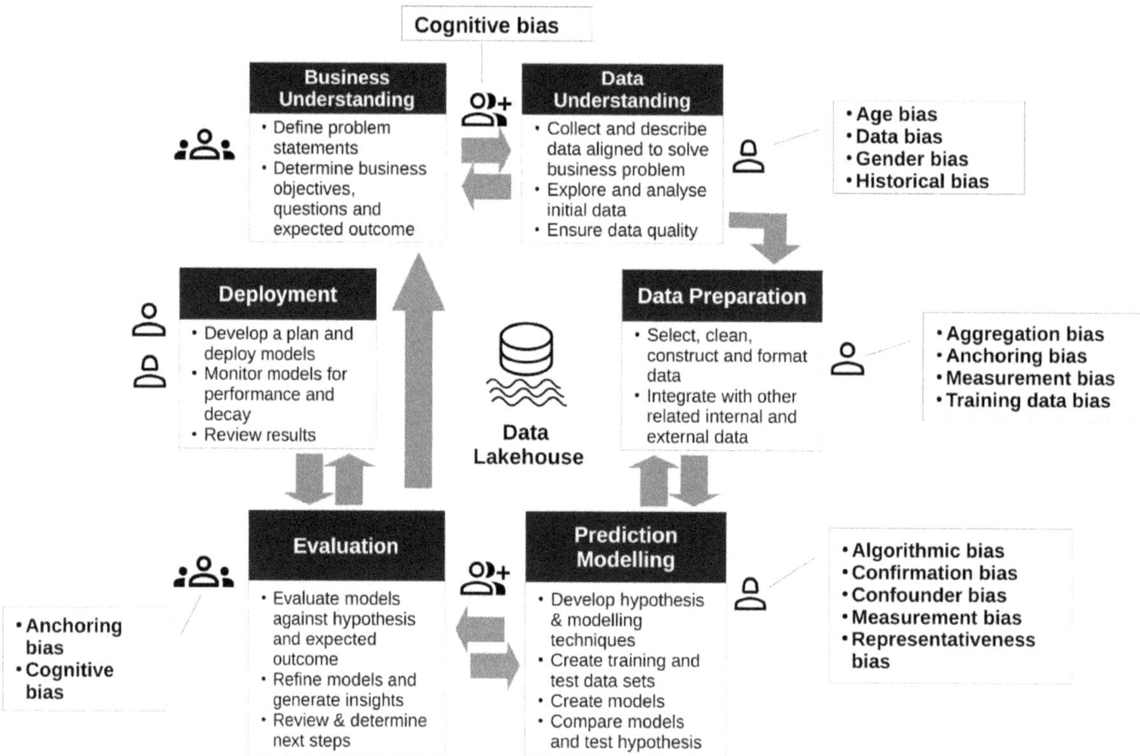

Figure 29: Biases in human in the loop (HITL) predictive model development process (Adapted from CRISP-DM)

Data Understanding: Data understanding is the process of analysing business problem statements, business questions related to the problem, and the required data that seeks to find answers to achieve business outcome. Data can be considered as the basis for creating information and knowledge. In contrast to data, information refers to a context created and communicated by the interpretation of data that has meaning to the recipient and can be understood by the recipient. This is an interactive process used by the data scientist that allows for adjustment of data aligned with business problems to be solved which will need to be made to the model depending on data availability, accessibility, and quality. It comprises data collection, verification of their relevance to the business problem, description of the data and quality to assess fitness, accuracy and trustworthiness of data on which business decisions can be reliability made.

'Although AI/ML systems have the benefit of freeing humans from laborious tasks and are able to complete complex calculations more quickly, they are only as effective as the data on which they are trained'.[17]

A number of biases can develop during this step, such as 'Age bias', 'Gender bias', and 'Racial bias', where people from certain categories or groups are favoured or excluded, thereby unfairly disadvantaging them.

'Data bias' can occur when the methods or sources used to collect data introduce a skew resulting in datasets that are not representative of the business problems or objectives. Similarly, collection of historical data containing past prejudices can unduly impact prediction results, known as 'historical bias'.

Data Preparation: Data preparation consists of selecting, filtering, constructing and formatting quality data suitable for being used during prediction modelling. It also involves the integration of data from internal and external sources that are used during model development. A number of biases can develop during this step.

'Aggregation bias' occurs when simplifying the training process by aggregating data into common groups that may result in overlooking some hidden details embedded in the grain of data.

During data preparation and analyses the data scientist or data analyst may be prejudiced to consider a particular value for an unknown quantity before estimating that quantity which leads to 'Anchoring bias' that can result in an anomaly in the results.

'Measurement bias' happens when the data used to train prediction models inaccurately represents the business problem or over-emphasises certain variables in the data.

When selecting data for training the model 'Training bias' can occur leading to models that are incomplete and/or not representative of the business problem/objective.

Prediction Modelling: As part of prediction modelling, the data scientist develops hypotheses about situations, and potential outcomes from observations and analyses of their understanding of the business problems, knowledge and business questions. Through such analyses the data scientist is able to identify patterns, trends, outliers, correlations, or other types of relationships between variables. Then, they identify modelling techniques by selecting the most appropriate

17 Siddique, S., et al., op. cit.

and relevant algorithms from the various categories of advanced analytical algorithms that we discussed earlier in Table 3. This can also involve multiple algorithms used in sequence, known as multi-genre or multi-modal analytics, to test their hypotheses on the training datasets that they prepared. They then compare the strengths of the analytical models against each other as well as with their hypotheses that can uncover correlations to make judgement of the models' effectiveness. During this process step, biases can develop.

'Algorithmic bias' can provide a biased result or misinformation when the business problem or the business question is not completely specific or correct. This can originate from the values and assumptions of the data scientist using imperfect training data that is not aligned with the business problem/question.

'Confirmation bias' and 'Confounder bias' in a model are due to the preconceptions or beliefs or assumptions of the data scientist when designing, developing hypotheses and selecting the algorithms.

'Measurement bias' can occur when the metrics or methods used to measure the performance of the model are inappropriate, inaccurate, or incomplete.

Similar to 'Confounder bias', 'Representativeness bias' is a type of 'Cognitive bias' that occurs due to the brain's attempt to simplify information processing by mental shortcuts or heuristics to solve the business problem.

Evaluation: The evaluation step involves assessment of the performance of the machine learning model across different groups to identify any disparities in the predictions. This includes standard evaluation metrics, such as accuracy, precision, recall, F1-score, or area under curve (AUC), softmax, and so on, for each group. By comparing these metrics across groups, you can identify variations in performance.

Prediction can generate new insights that are used to improve business process and outcome that often results in increased revenue, reduced costs, improving cash flows and profitability. Hence, the evaluation step also includes the appraisal and measurement of improvements in business performance that correlates with effectiveness of prediction. Business performance is evaluated from decisions and actions that are supported by insights provided by the prediction models. This is how data analytics (AI/ML) creates business value for the enterprise.

This step also allows for the business to retire some models that are not found to be effective or create new or enhanced models that align with changing business needs.

'Anchoring bias', which we previously identified in the data preparation step, can also occur in this step due to prejudiced assumptions when people consider a particular value for an unknown quantity before estimating that quantity, e.g. comparing everything to the first piece of information received.

Again, 'Cognitive bias' can occur during evaluation when people process information and make judgements influenced by past experiences and preferences.

Deployment: This step in the modelling process involves deploying the trained and tested model into operation so that the model can be used for business applications involving prediction of probability of an outcome from unknown data.

During operation of the model, it is also important to measure and track model performance due to the fact that model decay can happen due to changing characteristics of the data on which the models were initially trained. Model decay in production can have severe consequences with impairment in decision-making due to the absence of most up-to-date learning from changing data that did not exist when training the initial model. This allows for refining and retraining the models with changing data profiles so that they continue to be relevant during business operation.

In the next chapter, we will discuss how Agentic AI applications, involving natural language processing (NLP), integrate GenAI and RAG to retrain the LLM using real-time data to automate decisions.

CHAPTER 13

Interdisciplinary framework for decision-making

Overview: In the light of the arguments in the previous chapters on biases and limitations of humans and AI technologies, this chapter will discuss how to align the right mix of resources that produce the optimum desired business outcome involving automation and autonomous decision-making. We will also discuss the process of knowledge formation, observational learning, judgement and decision-making to determine a strategy for investment in decisions relating to people and AI technologies.

Automation and autonomous decision-making with AI

AI applications have been used in organisations with various capabilities during the past with some AI applications acting as software agents that interact with customers and employees to provide support services. *Agentic AI* is an emerging technology that leverages prediction models using machine learning to create autonomous conversational AI agents that are capable of understanding natural language, interpret complex instructions, engage in meaningful conversations, learn dynamically from trial and error, solve problems, make autonomous decisions for improving business outcome in every interaction with customers, business partners and employees. It combines GenAI, LLMs, RAG, and MCP (Model Context Protocol) to analyse data, set goals, take actions, learn from feedback on actions taken, and may choose to revise its approach in response to changing conditions, still ensuring that planned actions remain relevant. In effect, a distinguishing capability of Agentic AI is that it can perform tasks autonomously with decreasing human supervision. Nevertheless, as we covered in Figure 29, the human in the loop (HITL) is required where business users/managers set policies and provide the goals for the AI agents, ensure governance and intervene when human judgement and oversight is required.

Unlike deductive reasoning oriented routine tasks involving rule-based applications, Agentic AI mimics human cognition and inductive reasoning for routine and non-routine tasks. It works by developing patterns and probabilities from vast amounts of historical and real-time data that it gathers to infer and understand the context which it uses to determine likely outcome that influences its decision for next set of actions. Due to this capability,

Agentic AI is being used to automate workflows as part of business process re-engineering in business applications that interact with customers, employees and business partners (e.g. Customer Relationship Management; Customer Service Management; Customer Experience Management) and operational applications (e.g. Field Service Management; Asset Management; Finance Planning and Management; Supply Chain and Logistics Management). Workflows are also the means by which many autonomous Agentic AIs work together in a consistent flow in business processes.

Figure 30 and Figure 31 illustrate customer complaint management process before and after transforming to Agentic AI based application. In general, customer complaint can involve a range of reasons requiring remediation and resolution. In banking, retail, telecom and utility industries, there is frequent occurrence of over-charging customers for the product/service delivered that results in customers complaining about their bill. This can be due to gaps in internal processes arising between product, pricing and finance departments. In telecom and utility industries, service interruptions can be caused by faults and failures in the networks that provide communication and energy services to customers. In telecom, this includes dropped calls when using mobile phones for calling service or low upload/download speed when using data services or undelivered messages when using short message service (SMS). In utility industry, power interruptions and power cuts can lead to customer complaints. In automobile, retail and telecom industries, device or equipment failures can be common during warranty or out of warranty periods.

Figure 30 illustrates a common set of actions that arise when a customer has complaint about billing, service interruption or device malfunction. It usually involves the customer calling a contact centre and speaks with a human agent. The human agent then creates a trouble ticket and notifies the customer remediation centre. The customer remediation centre will perform preliminary investigation and then request each of the concerned departments to take corrective action. This can include making billing adjustments by crediting the customer's account with a refund, or inspecting and fixing a device fault remotely or by arranging for courier delivery, or resolving a service issue either remotely, or by sending a field technician to the physical location where the fix is required. The end-to-end process may take many days to resolve due to hands offs and other delays potentially causing inconvenience and frustrations to the customer.

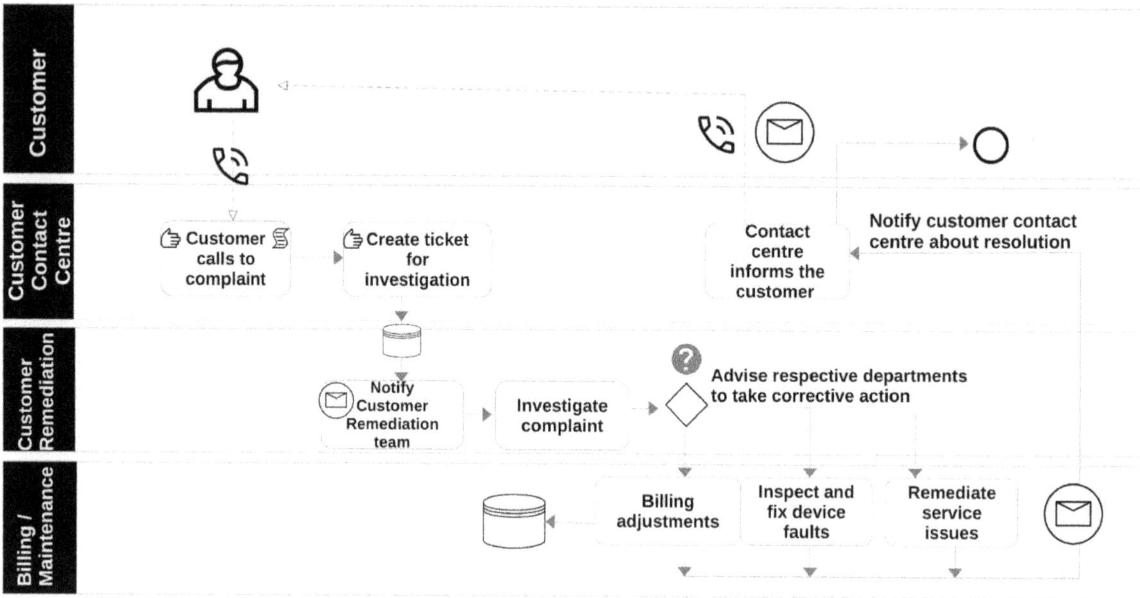

Figure 30: Customer complaint process before transforming to Agentic AI

Figure 31 contains an illustration of the transformed process with Agentic AI integrated into an existing CRM system. This involves a re-engineered process wherein the Agentic AI application consisting of multiple agents coordinate between themselves to execute different tasks in the chain driven by policies and guardrails set up by a human employee responsible for its administration. Some of these agents perform front-line customer facing tasks by using natural language processing capabilities of LLMs. Other agents retrieve external data from multiple sources to create context of the customer's complaint to determine if the complaint deals with over charged bill remediation or device fault or service interruption. Yet another agent may analyse the data using unsupervised machine language (ML) models such as K-Means, SVP (Support Vector Machines) or Random Forest or other techniques to develop patterns of the complaint and seek support from RAG to determine and infer reasons behind the complaint. Following this, one other agent will plan out the course of possible structured sequence of actions and make decision on most probable action that will likely resolve the complaint. The agent may also perform the action automatically by crediting the customer's account with a refund or remotely fixing faults where possible and then it will notify the employees involved about actions taken for audit purposes. Where automated action is not within its boundary, the agent will deflect to seek human intervention. It will use feedback from actions taken to learn and refine its prediction model.

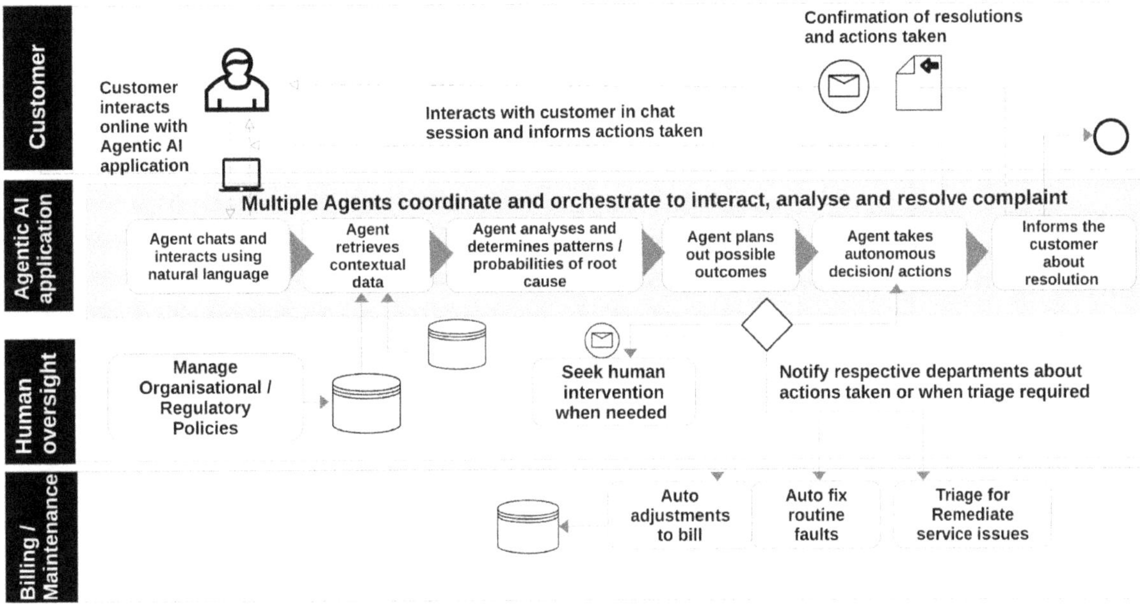

Figure 31: Customer complaint process after transforming to Agentic AI

Despite the unique and distinguishing capabilities of Agentic AI, it also has shortcomings due to its dependencies on pattern recognition and probabilities used by LLMs on which autonomous decisions are made. Popular LLMs also known to generate harmful content, produce questionable results, known as 'hallucinations'. To overcome such problems with general purpose LLMs, there is recognition in the industry that there is a need for domain specific LLMs, such as healthcare, banking, and telecom.

Agentic AI are also prone to potential biases, discrimination, and privacy violation in decision-making that could lead to unintended and harmful actions such as data breaches, violation of privacy and security that can impact the organisation in the form of damage to reputation or provision of misleading information to customers that can result in customer churn, financial loss or at worst legal actions. In order to prevent such incidents, many Agentic AI applications incorporate guardrails with the intention of continually monitoring its autonomous decision steps and actions to ensure compliance with organisation policies and prevailing government regulations.

Knowledge, judgement and decision-making

Performance of tasks in work activities require knowledge and skills (i.e. cognitive capabilities). Information gives an employee the ability to change their understanding and enhance/provide new knowledge when they perceive the need for information due to their current state of limited knowledge to fill the gap. They can identify patterns, trends, outliers, correlations, or other types of relationships between variables represented in the data by means of observation and analyses of data. These analyses may include predictive modelling with formal or informal formulations of hypotheses of their belief about situations and potential outcomes, which they can further analyse to uncover potential causal relationships.

Zaimoglu et al.[18] say, 'Human reasoning is the bridge that connects analysis of available data and information, to decisions about potential actions and their links to desired outcomes'.

However, since the employee is uncertain about decision outcome, they take risks by resorting to probability estimates of potential outcomes based on their current belief and hypotheses. As we discussed in Chapter 12 in the section 'Human cognition', predictions improve decisions but are imperfect and hence, there is a role for human judgement in adjusting decisions based on rewards from prior beliefs, potential rewards from new opportunity, prediction accuracy and other indescribable factors which are unique human capability (i.e. cognition, intuition, analogies, etc.).[19] But, they will likely stick to their current belief until new evidence accumulating against their belief passes a threshold that they set, at which point they may update their belief.

Figure 32 illustrates the above discussion using Bayes' Rule that shows the process of how someone updates their belief with new learning and knowledge after verifying their hypothesis against new evidence. Bayes' theorem or Bayes' Rule is named after Thomas Bayes, an 18th century English theologian and mathematician which allows one to find the probability of a cause given its effect. Posterior probability is the updated probability of an outcome after the new evidence is considered. Prior probability is the probability before the evidence is considered. Likelihood indicates probability of the evidence if the belief is true and marginal probability is probability of the evidence under any circumstance. Final decisions may still differ after considering the underlying drivers of the costs of prediction, the risks and rewards under the influence of intuitive aspects of expert decision-making such as intuitive human judgement, values and ethics. The feedback process helps in evaluating effectiveness of judgement and decision success.

---

18 Abdullah Kaan Zaimoglu, Lorien Pratt, Brian Fisher. (November 2023). 'Epistemological role of human reasoning in data-informed decision-making', *Frontiers in Communication,* 23 November 2023. https://doi.org/10.3389/fcomm.2023.1250301%20

19 Agrawal, Gans, and Goldfarb, op. cit.

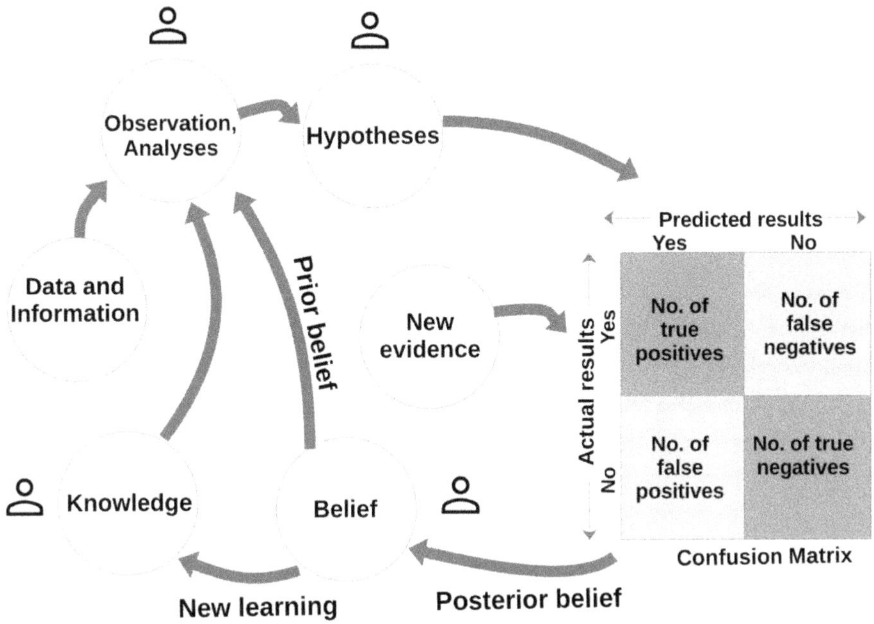

Figure 32: Bayes' Rule belief update

Human judgement involves weighing the risks and rewards of an action before making a decision. When people are unsure about the expected outcome from an action, they consider the risks of an action by evaluating the probability of the risk and associated reward when they make a judgement about the situation. According to behavioural economists and psychologists,[20] human judgement is flawed/biased, some of which we will review below in the context of how it affects decision-making. Following this, we will consider situations that incline towards automated decision-making by AI and those circumstances when augmenting human decision-making with AI will be favourable.

Limitations of human judgement include bounded rationality,[21] that argues that people must make decisions under the limits of cognition, information and time. This may be linked

20 Agrawal, Gans, and Goldfarb, op. cit.

21 Stuart Mills, David A. Spencer. (Dec 2024). 'Efficient Inefficiency: Organisational challenges of realising economic gains from AI', *Journal of Business Research* 189, 11512, 3 December 2024. https://doi.org/10.1016/j.jbusres.2024.115128

to the complexity of the decision being made under time constraints or as the number of decisions rises or when multi-tasking which limits how information can be cognitively integrated into the decision-making process.

According to Mills et al.[22] '*Boundedly rational decision makers engage in adaptive behaviours to overcome these challenges. These may include simplifying decisions and using mental shortcuts or heuristics. Actions taken are "satisficing" or "good enough" that allow a person to respond to and resolve whatever decision is facing them, while still making an acceptable choice given the limitations surrounding the decision*'.

Decision-makers might subconsciously favour some specific information that confirms their preexisting thought pattern, which is referred to as 'Confirmation bias'. Human judgement can be affected due to span of attention issues in judgement and paying attention to a larger number of factors may limit their effectiveness in judgement.

According to Daniel Kahneman,[23] losses loom larger than gains in human decision-making. He says people focus more on the risk of losing their possession than they focus on absolute benefits and the estimation of subjective probabilities is severely biased by 'Anchoring bias'. This is because the decision weights that people assign to outcomes are not identical to the probabilities of these outcomes. Improbable outcomes are over-weighted and outcomes that are almost certain are under-weighted relative to actual certainty.

Despite the many shortcomings of human decision-making discussed above, areas that currently remain more suited to human decision-making include social interaction, conflict management, teamwork, consensus seeking and ethical judgement.[24] Humans possess unique capabilities in making assessment of rewards which AI is unable to assess. Human cognition is flexible, but of limited capacity while the machine is rigid, but has ample capacity. AI mainly uses algorithms to learn from iterative cycles of data analyses to find insightful information without deterministic programming. However, limitations in AI occur because such machines have no conscious states, no mind, no subjective awareness, and are often disconnected from the execution and outcomes of decisions. AI detects patterns from vast amounts of data that are invisible to human eyes. Humans have information through personal interaction which is not observable by AI.

Considering the pros and cons of humans and AI, collaboration between AI and human is suggested to be more beneficial. When prediction is cheap, the volume of decision-making increases and risk of predictions is low then, AI will be suitable for automated and

22 Ibid.

23 Kahneman, D. (2011). *Thinking, fast and slow,* Farrar, Straus and Giroux, New York.

24 Dawid Booyse, Caren Brenda Scheepers. (27 June 2023). 'Barriers to adopting automated organisational decision-making through the use of artificial intelligence'. https://www.emerald.com/insight/content/doi/10.1108/mrr-09-2021-0701/full/html%20

autonomous decision-making. When the risk and rewards are high, human judgement will be called for to make adjustments to predictions and final decision-making. Therefore, strategic decision-making requires a combination of both human and machine intelligence to deliver augmented solutions that can help to address decision-making within complex situations.

In the light of the above discussions, Figure 33 is a revised version of Figure 28 which now maps task characteristics with human and AI/ML capabilities along with risk and reward probabilities. Routine tasks that involve low risks or low risks with high rewards are candidates for substituting human capabilities with AI/ML. Non-routine tasks with high risks or high rewards are best suited to humans to perform that can be augmented with AI/ML for enhanced performance due it its capability of understanding context, adapting to new information, and collaborating with humans to solve complex challenges.

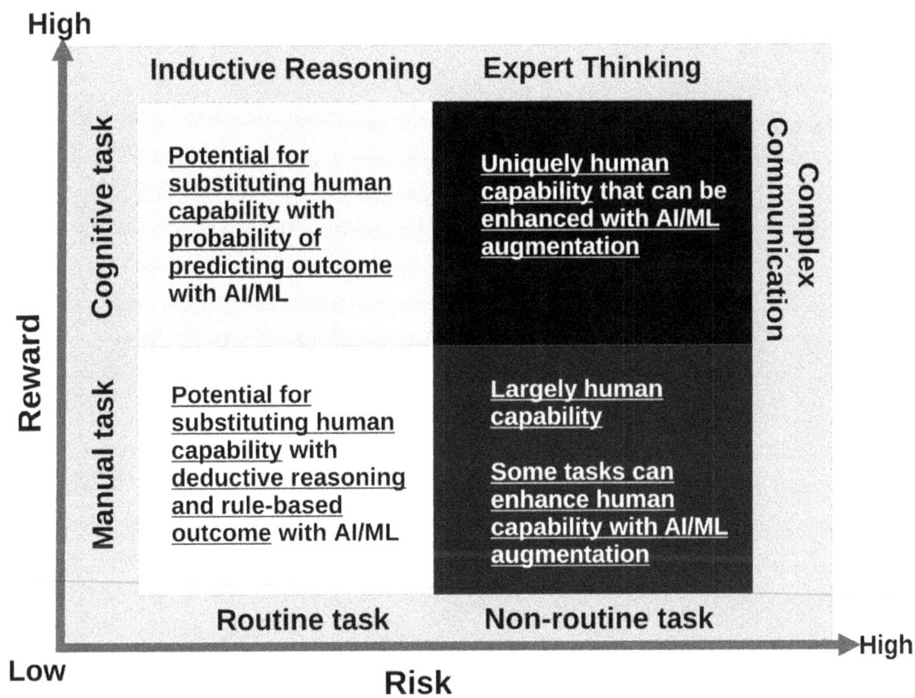

Figure 33: Human and AI/ML capabilities mapped with task characteristics along with risk and reward probabilities

Nevertheless, when augmenting with AI/ML, it is important to take measures to address AI/ ML bias by developing more transparent AI applications with audit trails and explanations of their decision-making steps which will enable users and managers to understand and therefore, trust the intelligent agents. These measures include governance of data and prediction models.

Data governance comprises oversight to ensure that data normalisation, feature scaling, quality assurance, data aggregation, or augmentation during 'Data Understanding' and 'Data Preparation' stages do not introduce biases. Model governance ensures validation and cross-validation of model performance measures are consistent between different groups and are sensitive to thresholds and decision boundaries. Periodic reviews by independent domain experts or domain specialists are also essential to ensure models' prediction accuracy and associated decisions are consistent with expected outcome.

CHAPTER 14

Techno-human dynamics in decision-making

Overview: The previous chapters of this book explored the aspects of human decision-making, unique cognitive capabilities of knowledge workers and the emerging trend of using artificial intelligence (AI) technologies for autonomous decision-making in the organisation. These techno-human dynamics have created tension in the minds of executives that lead to the dilemma of investing in human versus AI in the organisation.

The focus of this chapter is to synthesise the preceding discussions through the perspective of an academic clinical psychologist and researcher whose expertise lies in human cognition and the biases that shape information processing. It examines how the learning architectures of artificial intelligence, together with its predictive capabilities and cognitive constraints intersect with enduring psychological principles of attention, emotion, and reasoning. This integration provides a deeper understanding of how the science of human thought can illuminate both the potential and the risks inherent in our evolving partnership with intelligent technologies.

The insights developed in Chapter 13, combined with the recommendations outlined in this chapter, aim to establish a framework for designing an enterprise-level strategy that navigates the challenge of optimally allocating investment between human and artificial intelligence to strengthen organisational judgement and decision-making.

Rethinking decision-making in the age of algorithms

For centuries, the human mind has served as the most sophisticated decision-making mechanism on earth, governed by an intricate interplay of logic, memory, emotion, and intuition. These cognitive faculties have shaped civilisations, advanced science and technology, and inspired artistic innovation. Today, however, we face a new counterpart in decision-making: the machine.

Artificial intelligence, long associated with speculative narrative, has now assumed an operational role across diverse enterprise domains, including customer engagement, clinical diagnosis, financial forecasting, and creative production. Its accelerating integration into organisational processes raises a critical strategic question: in shaping the decisions that determine organisational direction, societal outcomes, and the future contours of human activity, should we assign greater confidence to human judgement or to algorithmic precision?

The emerging tension between human and artificial intelligence reaches beyond technological capability. It brings into focus not just philosophical, but also behavioural–science considerations concerning the nature, value, and inherent limitations of human reasoning. Addressing this tension requires a more systematic and objective examination of human intelligence itself, including whether distinct cognitive attributes continue to differentiate human judgement and decision-making in an era increasingly influenced by data-driven and algorithmic systems.

Reassessing human judgement in an era of intelligent systems

The enduring strength of human intelligence lies in its flexibility and depth, and its capacity to interpret context. It thrives where information alone is insufficient, be it in moral dilemmas, creative breakthroughs, and social complexity. It is shaped as much by experience as by reasoning, and as much by intuition as by evidence. This interpretive quality allows humans to navigate ambiguity, discern intention, and balance competing values. Intelligence, in this sense, is not merely a cognitive process, but a living synthesis of memory, emotion, and foresight, our ability to imagine, to feel, and to act purposefully in an ever-changing world.

At the core of this interpretive capacity lies intuition, often described as a tacit, rapid form of knowing that draws on accumulated experience. Intuition transforms prior learning into immediate insight. For example, a senior emergency physician detecting a subtle shift in a patient's emotional expression or narrative coherence that signals underlying deterioration before clinical measures reveal it, or a policy adviser anticipating significant public resistance to a regulatory reform well before formal consultation data indicate emerging tensions. In both domains, experienced professionals recognise nuanced cues, whether a momentary hesitation that reveals unspoken distress or a change in stakeholder engagement that hints at impending fragmentation that signal rising risk. As Gary Klein's Recognition-Primed Decision (RPD)[25] model demonstrates, intuition reflects experience operating at speed, knowledge compressed into an instant of recognition. It becomes especially valuable in environments characterised by uncertainty, incomplete information, and rapidly evolving conditions, where deliberative analysis alone may be insufficient. Many clinicians, policy leaders, and enterprise leaders attribute their most consequential decisions not solely to formal analytics or procedural frameworks but to these intuitive judgements, where expertise, pattern recognition, and contextual awareness converge.

25 G.A. Klein. (1993). 'A recognition-primed decision (RPD) model of rapid decision making'. In G.A. Klein, J. Orasanu, R. Calderwood, & C.E. Zsambok (Eds), *Decision making in action: Models and methods* (pp. 138–147). Ablex Publishing. https://psycnet.apa.org/record/1993-97634-006

Human intelligence as information-processing framework

Human intelligence excels in creative and context-dependent reasoning, but its performance is characterised by variability rather than uniform consistency. Our decisions are coloured by mood, fatigue, social influence, and personal history. The same person may make different choices in the morning and at night, after coffee or after conflict. For a moment, let us consider human intelligence through the lens of an information-processing framework, not as cold computation, but as a living system where memory, cognition, emotion, and meaning converge to shape our decision-making. Like computers, we receive input, encode and process information, and produce outputs, judgements, decisions, and behaviours. But unlike machines, our processing is not purely computational. It is influenced by memory, cognitive state, emotion, motivation, and context, layers of complexity that enrich human intelligence, but also expose it to information-processing biases that can distort perception and judgement. While Chapter 12 described information-processing biases during the data analytics (AI/ML) model development process, the following paragraphs illustrate the biases from the perspective of human psychological processes which has significant implications for management judgement and decision-making at the enterprise level.

For example, at the input stage, we perceive information through our senses, but perception itself is selective. We don't absorb reality as it is, we absorb it as we notice it. Attention filters and prior experiences shape what enters consciousness. Two people may observe the same event and encode entirely different interpretations. During encoding and processing, the brain organises, interprets, and stores information. Here, cognitive biases typically creep in, shortcuts that help us make fast judgements but can distort truth. Affective (emotional) biases and motivation also influence which data we prioritise and how we interpret them. Finally, in the output stage, we act or decide. Our choices are shaped not only by logic but also by emotional states, moral values, and social expectations. The result is an intelligence that is as affective as it is cognitive, a system built not merely for efficiency, but for meaning. This makes human decision-making, for all its sophistication, remarkably preferential and fragile.

Psychologists Daniel Kahneman and Amos Tversky[26] demonstrated that even the most rational individuals are predictably irrational, their judgements subtly shaped by cognitive biases that often operate outside conscious awareness. We seek evidence that confirms existing beliefs, mistaking familiarity for truth, and fear what is most vivid in memory rather than what is statistically probable. We are swayed by first impressions, anchored by initial

26 Tversky, A., and Kahneman, D. (1974). 'Judgement under Uncertainty: Heuristics and Biases', *Science*, 185(4157), 1124–1131.

information, and influenced by recent experiences, frequently neglecting the broader temporal or contextual frame.

Human memory itself is not a fixed record but a reconstructive process, dynamic, selective, and susceptible to distortion.[27] Phenomena such as the misinformation effect illustrate how memories are continually revised, integrating new information or affective states into narratives that may deviate from objective reality. We persist in failing endeavours because prior investment biases perception, and we anchor judgements of value to the earliest figures or ideas encountered. These tendencies, including confirmation bias, availability bias, anchoring bias, memory bias, primacy and recency effects, and the sunk cost fallacy are not defects of intelligence, but adaptive heuristics, evolved to facilitate rapid decision-making under uncertainty. Yet they reveal that human reasoning, while extraordinary, is neither purely logical nor objective. It is but an intricate preferential synthesis of cognition, emotion, and context, a system perpetually negotiating between insight and distortion. Notwithstanding these limitations, individual differences in memory, cognition, and perception can serve as a valuable resource in collective decision-making. For example, within enterprise contexts, such cognitive diversity enhances feedback quality and contributes to more robust and well-reasoned group consensus.[28]

The unblinking eye: Optimisation beyond human reach

Throughout history, humanity has developed tools to extend its perceptual and cognitive boundaries, the telescope to magnify the cosmos, the microscope to unveil the microscopic, and the computer to augment calculation. Artificial Intelligence (AI) represents the most recent and transformative of these extensions, redefining the nature of decision-making itself. Unlike human cognition, constrained by biological limits of memory and processing capacity, AI systems can ingest, organise, and analyse vast quantities of data with exceptional speed and precision. Their capacity to identify correlations and anomalies across multi-modal datasets, customers' perceptual experience, financial transactions, medical scans and environmental sensors enable insights that surpass the scope of human analyses. In domains such as fraud detection and diagnostic imaging, AI's ability to recognise patterns across millions of data points in real time has produced measurable improvements in accuracy and efficiency, establishing it as an indispensable component of contemporary decision-making architectures.[29]

27 Loftus, E. F. (2005). Planting misinformation in the human mind: A 30-year investigation of the malleability of memory. Learning & Memory, 12(4), 361–366.

28 Woolley et al. (2010). Evidence for a collective intelligence factor in groups. *Science*

29 Topol, E. J. (2019). High-performance medicine: The convergence of human and artificial intelligence. *Nature Medicine*, 25(1), 44–56.

AI's analytical power lies in its capacity to detect complex, non-linear relationships that elude human reasoning. While human cognition often proceeds through linear inference or narrative-based logic, AI operates within multidimensional networks of probability, identifying associations without presupposed hypotheses. In medical research, for instance, machine-learning models have uncovered correlations between retinal features and cardiovascular disease risk, connections that no clinician had previously theorised.[30] Similarly, in industrial maintenance, predictive algorithms discern micro-level deviations in vibration or temperature that precede mechanical failure, allowing for pre-emptive intervention.[31] Such capabilities demonstrate how AI extends the epistemic horizon of data-driven judgement, converting unstructured information into predictive knowledge through empirical computation rather than intuitive reasoning. Equally significant is AI's procedural consistency and resistance to cognitive distortion. Human judgement is inherently variable, shaped by emotion, fatigue, and contextual bias. AI systems, by contrast, apply fixed parameters uniformly across cases, offering a degree of reliability and standardisation that human cognition cannot replicate.

Algorithmic bias: The not-so-subtle threat

There is increasing awareness of the integrity of the data from which AI learns. When historical data reflect structural inequities or sampling biases, these distortions can be replicated or amplified in algorithmic outputs. Consequently, AI's objectivity is conditional rather than absolute. It reflects not inherent neutrality but the epistemic quality of its inputs and design. Nonetheless, when rigorously audited and ethically governed, AI can function as an extension of human judgement rather than its replacement, a methodological instrument capable of transcending human cognitive and physical limitations and enhancing the precision, scale, and fairness of data-informed decision-making.

While these capabilities illustrate the transformative potential of AI in augmenting human decision-making, they also reveal a fundamental tension at the core of algorithmic judgement. The very attributes that make AI powerful, its reliance on data, statistical inference, and procedural consistency, also delimit its understanding of context, meaning, and moral nuance. Algorithms excel at correlation but remain indifferent to consequence. Thus, as AI becomes increasingly embedded in systems of governance, healthcare, and commerce, questions

30 Poplin, R., Varadarajan, A. V., Blumer, K., Liu, Y., McConnell, M. V., Corrado, G. S., Peng, L., & Webster, D. R. (2018). Prediction of cardiovascular risk factors from retinal fundus photographs via deep learning. *Nature Biomedical Engineering*, 2(3), 158–164.

31 Carvalho, T. P., Soares, F. A. A. M. N., Vita, R., Francisco, R. da P., Basto, J. P., & Alcalá, S. G. S. (2019). A systematic literature review of machine learning methods applied to predictive maintenance. *Computers & Industrial Engineering*, 137, 106024.

arise not merely about its accuracy, but about its alignment with human values and ethical reasoning. To comprehend the limits of algorithmic intelligence, it is necessary to examine where computational precision ends and interpretive understanding begins.

AI's strength lies in its procedural consistency and resistance to the cognitive distortions that were outlined in the previous section that affect human judgement. Unlike humans, whose decisions are influenced by fatigue, mood, or social context, AI applies the same logic uniformly across cases, offering reliability and standardisation at scale. However, this consistency is not synonymous with objectivity. It is entirely contingent on the integrity of the data from which the algorithm learns. For example, when training data reflect historical inequities such as gender imbalances, racial disparities, or socioeconomic exclusions, those distortions become encoded within the model itself, producing systematic and self-reinforcing bias.

This phenomenon, known as algorithmic bias, represents one of the most insidious risks of enterprise AI besides the risk of model decay[32] that can occur when real-world data deviates from the data on which the model was initially trained. These systems do not discriminate by intent but by limitation, reproducing historical patterns rather than normative ideals. Hiring models may favour male candidates in organisations with gender-skewed histories, and lending models trained on legacy credit data can inadvertently disadvantage minority applicants. The risk is amplified by scale: a human error is isolated, but a biased algorithm can replicate the same flaw across millions of decisions before it is detected.

These challenges expose a fundamental paradox in algorithmic judgement. The very features that make AI powerful, its reliance on data, statistical inference, and procedural consistency, also limit its ability to interpret context, meaning, and moral nuance. Algorithms optimise, they do not understand. They correlate without comprehending, and they predict without empathising. This limitation becomes especially salient when decision-making relies on subtle interpersonal cues and contextual insight, areas in which human clinicians excel, and AI systems may falter. As AI systems become increasingly embedded in governance, healthcare, commerce and enterprises, the question is no longer merely whether they are accurate, but whether they are aligned with human values and ethical principles.[33] Addressing this tension demands regular bias audits and sustained human oversight.[34] Only under such

32 Gama, J., Žliobaitė, I., Bifet, A., Pechenizkiy, M., & Bouchachia, A. (2014). A survey on concept drift adaptation. ACM Computing Surveys, 46(4), Article 44, 1–37. https://doi.org/10.1145/2523813

33 Mittelstadt, B. D., et al. (2016). 'The Ethics of Algorithms: Mapping the Debate', Big Data & Society, 3(2). https://doi.org/10.1177/2053951716679679%20

34 National Institute of Standards and Technology. (2023). Artificial intelligence risk management framework (AI RMF 1.0) (NIST AI 100-1). U.S. Department of Commerce.

rigorous governance, can AI function as an extension of human judgement rather than its distortion.

Over-reliance on AI and the automation trap

One of the most subtle yet dangerous risks in AI adoption is over-reliance, the human tendency to defer uncritically to algorithmic authority. Known as automation bias,[35] it arises when decision-makers assume the precision of AI equates to accuracy.[36] Overconfidence in algorithmic outputs has repeatedly demonstrated that blind trust can magnify, not mitigate, systemic risk. When human oversight recedes, errors compound silently and at scale.

Beyond operational risk, automation complacency erodes human expertise. As organisations increasingly depend on AI for recruitment, lending, and evaluation, the subtle arts of decision-making and contextual reasoning risk atrophy. Over time, human decision-makers become auditors of machine logic rather than active interpreters of meaning. The resulting systems, while efficient, grow brittle, optimised for routine but ill-equipped for anomaly or crisis. True intelligence in enterprise therefore demands balance. Algorithms should inform human choice, not replace it.

Finally, the future of enterprise intelligence depends on cultivating symbiosis rather than substitution. AI's precision and scale can enhance decision-making only when tempered by human contextual awareness, ethical reasoning, and emotional intelligence. The next evolution of the enterprise mind is not artificial but augmented, a partnership in which machines handle complexity, and humans provide conscience.[37] The challenge is not to trust AI more, but to trust it wisely, ensuring that as our systems grow smarter, our judgement grows deeper.

Transition: Toward a new partnership

The lesson from the automation trap is clear: Intelligence, whether human or artificial, reaches its highest potential only through integration. Neither human intelligence nor computation alone can navigate the complexity of modern decision environments. The future of enterprise and societal judgement lies in hybrid cognition, a deliberate partnership

---

35 Romeo, G., Conti, D. (2025). 'Exploring automation bias in human–AI collaboration: a review and implications for explainable AI', *AI & Soc* (2025). https://doi.org/10.1007/s00146-025-02422-7

36 Rosbach, E., Ganz, J., Ammeling, J., Riener, A., Aubreville, M. (2025), Automation Bias in AI-assisted Medical Decision-making under Time Pressure in Computational Pathology, In: Palm, C., et al. *Bildverarbeitung für die Medizin* 2025. BVM 2025. Informatik aktuell. Springer Vieweg, Wiesbaden. https://doi.org/10.1007/978-3-658-47422-5_27

37 Wilson & Daugherty (2018). Collaborative Intelligence: Humans and AI Are Joining Forces. *Harvard Business Review.*

between human discernment and machine precision. In this emerging model, humans are not passive overseers of automation, but active collaborators within adaptive systems that learn and reason alongside them. Here, AI becomes not a rival intellect but a cognitive extension amplifying human insight while remaining grounded in ethical, emotional, and contextual awareness.

Part VIII

Approaches for measurement and management of data analytics (AI/ML) value

Part VIII recognises many of the shortcomings discussed in the previous chapters and proposes Data Analytics Value Framework (DAVF). DAVF extends the value measurement principles covered in Chapters 6, 7 and 8, for effective measurement and management of long-term value. These include principles, processes, methods, techniques, and tools for data analytics value assessment/realisation (DAVA/DAVR), value estimation (DAVE) and value optimisation (DAVO) covered in their respective chapters.

CHAPTER 15

A holistic approach for measuring data analytics value

Overview: In the light of the discussions in the previous chapters, this chapter of the book will review and summarise the main challenges. We also argue for taking holistic approaches to measuring, quantifying and attributing the long-term business value of data analytics (AI/ML).

'Spaghetti web' of technologies and tools

A glance at the technical architecture of the IT environment of a large bank, telecommunication or retail organisation can quickly reveal the complexities of the interconnections between them that can best be described as a *'spaghetti web'*! The transformation of the IT and IS landscape in concert with its evolution over the different computing eras that have been instrumental in ever-growing applications, systems, data and computing equipment. As can be seen in Figure 34, in recent years the IT and data analytics environment has become even more complex with varieties of technology and tools supplied by various vendors. Just a glance at the Hugging Face website reveals that there are over 2 million large language models[38] and over 500,000 data sets.[39]

In any data centre of a large organisation there are multiple types of hardware infrastructures ranging from multi-cloud, on-prem processors, storage and network equipment, different types of database technologies (e.g. data marts, data warehouses, data lakehouses, NoSQL databases, vector databases), multiple categories of data management, governance, extract-transform-load tools, data quality and observability software, business intelligence, visualisation, reporting tools as well as computer vision, speech recognition, NLP and AI/ML tools.

Many of these integrate with enterprise application software such as ERP, CRM and office automation/collaboration tools. Home grown systems, as a result of the Open Source movement, have also contributed to the complexity.

38 https://huggingface.co/models

39 https://huggingface.co/datasets

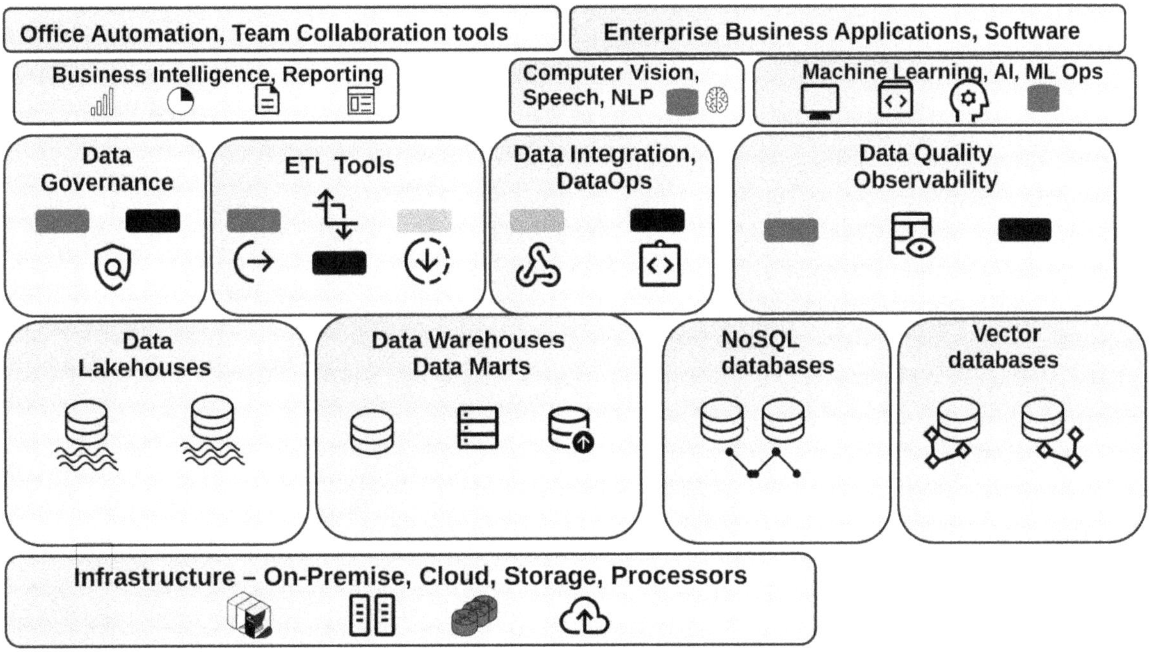

Figure 34: Complexities of data analytics environment calls for holistic measurement of its value

Each of these technologies and tools have differing costs components and their value contributions to the business units and the enterprise, and in differing proportions of value for their investment. Any technology solutions to a business problem generally involves a combination of technology, tools, data and analytics. Therefore, the total cost of the solution, and the value contributions of each technology, tool, data, analytics and human skills can best be considered as the portfolio of cost and value. While this resembles the Bill of Materials (BoM) of components that go into a finished product in an assembly line in manufacturing, the characteristics and attributes of data analytics are not comparable. In short, physical assets depreciate over time with usage, while data analytics constantly appreciate with enrichment and usage, and grow in recurring value over time.

Evolution of data variety, volume and schema

Adding to the complication of the IT and IS environment is the fact that there has been increasing variety, volume and speed at which data has been growing in the organisation from megabyte size to petabyte size during each stage of their evolution (see Figure 35). It is also noticeable that the format of data has undergone significant mutations from stable and well-defined schema, to evolving schema, no schema, and unstructured data formats. As a consequence of this mutation, there are added intricacies in formulating metadata management in data governance. As the volume of data has grown, it has also altered the cost and value economics. Ever since the evolution of the Internet, growth in volume has stunted value by increasing the cost of data along with redundant and unusable volumes of data (e.g. repeated markups and tags in web logs, HTML, XML and JSON).

Challenges in linking data analytics investments to value

In conjunction with and as a result of the data complexities we discussed above, there have been growing regulations requiring compliance in the area of data security, privacy and trustworthiness of AI that impacted many industries. Some industries have had increasing scrutiny and regulation. The financial services industry has seen anti-money laundering, capital adequacy and liquidity regulations and the telecom industry had law enforcement and data protection regulations that called for retention and provision of logs of calls (Call Detail Record), messages, web browsing, data upload and download from any device to be reported to the regulators. Healthcare regulations on data privacy had also escalated in recent years.

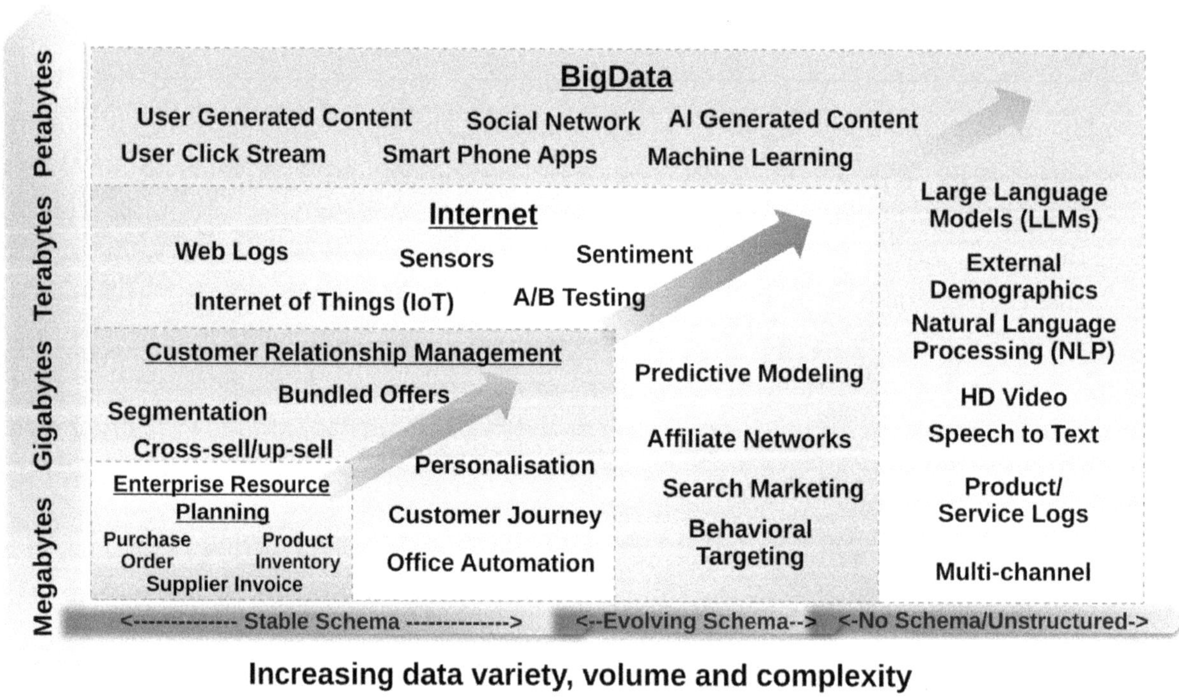

Figure 35: Increasing complexity with evolution of data variety and volume

As a result of this regulatory evolution, as well as data growth, there have been growing problems in linking the value contributions to its investment due to the mandated requirements that forced organisations to incur costs and expenses without considerations for returns for these investments.

Additionally, there are many internal business functions and activities in the organisation that face difficulties in measuring the ROI of data analytics. Take the CFO office for instance, where the general ledger has been the main source of corporate financial planning, budgeting, forecasting and reporting of profit and loss (P&L), cash flows and balance sheet (i.e. assets and liabilities). Large organisations typically have multiple general ledger systems spanning their business functions, divisions and regions, each with their own local reporting practices and regulations with differing general ledgers that make it difficult to create a single view of the corporation in time for reporting. Because the general ledgers are a repository of transactions that are dynamically and constantly changing it is difficult to report on the state of the corporation with certainty for the stakeholders and regulators.

Hence, the financial reporting standards and regulators require the organisation to earmark a day on which all financial transactions of the enterprise are synchronised, reconciled and complete for periodic reporting; this is called 'closing the books'. Timely closing of the books is measured as d+1, d+2, d+3, etc., meaning the financial books are closed one, two or three days after the set cut-off date. Most financial controllers aim for d+1 in closing the books. Any delay in closing the books affects the reporting of accurate state of the business that enables organisations to make trustworthy decisions. The general ledger has other limitations that prevent it from being used for analyses such as,

'Sales Orders versus Invoices per Customer'

'Discounts per Product/Service per Customer'

'What activities are the main cause of the distribution cost differences'

'How revenue has been lost due to leakage', etc.

In order to meet these requirements, the CFO's office invests in integrating the multiple general ledgers in the data analytics ecosystem. A simple question, such as *'Is there ROI for closing the books on time? If so, what is the ROI for closing the books on time?'* is very difficult to answer, mainly because this is often considered as the cost of doing business or an overhead. In such cases the financial controllers make their own judgement of the ROI based on the organisational strategy, objective and their key performance measures. Their judgement and perceptions can differ between organisations.

Here are two differing perceptions that were encountered during the consulting practice.

During the meeting with the director of finance and accounting of a large telecommunications company the comments and question to him went along this line,

'In the interest of closing the books on time, accountants generally make estimates when actual transactions don't arrive on time and they reverse the transaction by means of journal entries in the general ledger on a date after closing books when the actual transactions arrive. Given this scenario what is the ROI or benefit of closing the books on time, within say d+1 or d+2, with actual transactions in the general ledger?'

His response was,

'During the period when the actual transactions are not known our estimates can be wrong and we may be running blind in our decision!'

Following this he added,

'I estimate the ROI of data analytics that allows us to close the books on time is 1 week of the company's revenue.'

When making the same comments and asking the same question to the head of financial accounting of a tier-1 bank her response was:

'Yes, you are right we make estimates and reverse the transaction after the month end closing, but it is very important to close the books with actual transactions for half-yearly and annual financial reporting because we announce these results to the market.'

Her follow-on response was:

'If we report incorrect results, it will affect share prices and shareholder value which will have legal repercussions. The ROI of data analytics is in avoiding such backlashes.'

In this part of the book, the data analytics value framework provides a process and method for measuring and quantifying data analytics value for such situations which is referred as 'notional value' determined by 'probabilistic' estimate of value by understanding the perception and judgement criteria of the stakeholder. On the other hand, where the data analytics value can be easily linked to the investment, such as in marketing campaigns, it is referred to as 'deterministic' or 'tangible value'.

Business and technology organisational divide

Besides the IT and IS technology evolution, there has been divide in the organisation with the rise of the roles of Chief Information Officer (CIO) during the Distributed Computing Era and Chief Data Officer (CDO) during the Internet Era respectively. This has contributed to the stark division in the organisation between business management and technology management (see Figure 36). The CIO and CDO focus on Total Cost of Ownership (TCO) while C-suites in the business divisions focus on Return on Investments (ROI).

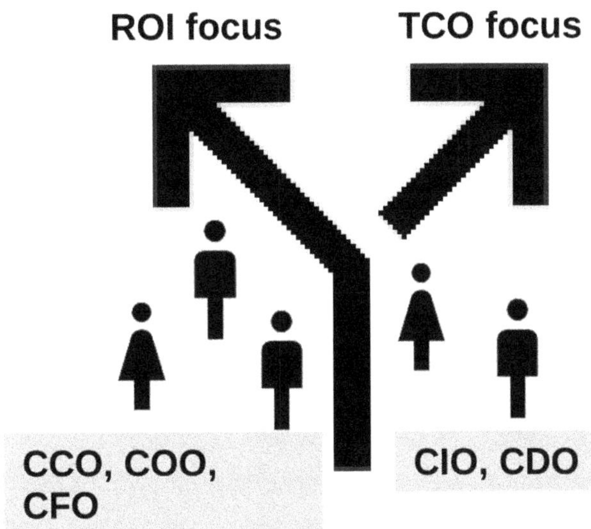

Figure 36: Business and technology divide leads to differing value perceptions and focus

A survey of CEOs, undertaken by the Gartner Group[40] during the early stages of digital transformation, confirmed what was known all along, that the IT department thinks of innovation in terms of technology rather than business alignment. Gartner reported that IT focuses on cost (easily measurable), while they ignore (or don't measure, at least) the thing that's most valuable to the business. The survey mentioned that CIOs have the opportunity to match the expectations of the CEOs, developing greater insight across all areas of the enterprise for enabling process efficiencies, customer insight and innovation, plus cost control. It went on to underline the need for a change to a more vision-led and inspirational leadership style ('vision first' rather than 'control first').

Transformation of a cost controller

It is a different story on the business side. In a customer meeting in a large organisation in New Zealand, discussing the need to use data and analytics to drive enterprise value, one of the participants from the customer side volunteered,

'We have four customer segments, nationally, that we have been using for a very long time with reasonable success but, it is limiting our growth. Strategically, we are now moving towards a regime of micro-segmentation with an objective of driving data-driven decisions for local area marketing. Our CFO is very interested to know how the Net Promoter Score (NPS) is improving in locations where we have made significant infrastructure capital investments.'

40 http://www.gartner.com

'Coming from the marketing guys, this kind of thing would have been a non-event for me. But, when discovering that this was their head of Finance talking, it came as a pleasant surprise. When I asked him what was behind the transformation of their traditional Finance department – better-known as the 'bean counters', he laughed, explaining that the costs involved in acquiring and retaining customers are affecting profitability. He said, as a consequence, and strategically, Finance is aligning itself with Marketing, Customer Service, and IT, to grow customer satisfaction, and improve margin and profitability. He added that he believed data and analytics are key factors in unlocking that growth.'

Although there are differing perceptions and judgements of value between the different stakeholders, it is important to take them into consideration when measuring and attributing data analytics value, with a central requirement that they are aligned with the organisation's strategy, objectives and KPIs of the responsible stakeholders. For example, as part of their mandate to manage cost of data centres and Cloud computing resources efficiently and effectively, the CIO and CDO focus on TCO by optimising costs of software licensing, software development and maintenance, computing servers/hardware (CPU, memory, storage), networks, and related infrastructure and facilities. On the other hand, stakeholders on the business side are responsible for revenue generation, gross margin, cash flows, profitability while also ensuring cost efficiencies. The data analytics value framework considers these along with other complexities which we will explore in the next chapter.

CHAPTER 16

Data analytics value framework

Overview: The previous chapter of the book laid out the need for taking holistic approaches to measuring, quantifying and attributing the long-term business value of data analytics (AI/ML). This chapter provides an overview of data analytics framework and describes the resources and tools to assist with the holistic approach in a pragmatic way.

An overview of Data Analytics Value Framework (DAVF)

The division of responsibilities and perceptions of value between the business and technology organisation that we reviewed in Chapter 15 has an additional level of complexities. For instance, the notion of data as a *resource-driven principle* wherein as part of their planning process the data managers in the CDO's office continually enhance and 'ring-fence' central repository of enterprise data with an expectation that it can be tapped at any time with the certainty of achieving a predictable value from it. So, the limits imposed on the organisation's data lakehouse ecosystem are based upon the data managers' own perception of their value as information.

Conversely, data as an investment-focused growth asset assumes a *perception-driven model* where the individual front-line employees' perception of the data widens because of their greater accessibility to the information across the enterprise. Hence, on the one hand, the value of information is limited by the boundaries drawn up by the data managers, and on the other hand, expanded by the perceptions of the business users and customer-facing employees having access to the enterprise network of data. So, information is the consequence of a complex psychological process that transforms perceived data into usable thought inputs by using the skills of employees for the benefit of improved customer communication and experience, and organisational performance. It not the data *per se*, but the time, utility and context of data which transforms it into information that enables actionable decisions, and hence value to the organisation.

Besides the shifts in the perception-driven model, the evolution of data analytics has given rise to Agentic AI. Due to its ability to mimic the cognitive skills for tasks traditionally performed by knowledge workers, Artificial Intelligence (AI), equipped with Machine Learning (ML) algorithms is top in the minds of executives in making investment decisions involving potential substitution of employees (i.e. real intelligence) in the organisation. But, data analytics on which AI applications rely as well as employees in the organisation are not treated as assets in the balance sheet, thereby blindsiding executives as to their true long-term value on which to make investment decisions. In the light of the dilemma faced by executives as to whether to invest in employees who possess real intelligence or to substitute them with AI technologies, the remaining chapters of this book offer strategies and tactics for effective measurement and management of data analytics investments as growth assets by avoiding the shortcomings in the balance sheet and applying the data analytics value framework (see Figure 37) described in the book. It is a far cry from the historical value engineering approach that has been focused on cost reduction.

Figure 37: Data analytics value framework

The foundational principles of the data analytics value framework are based on:

- strategic long-term value with considerations for unique characteristics of data that we discussed at length in Chapter 3, under the heading 'Assets: Physical versus data analytics (AI/ML)'

- aligning value to the perceptions of stakeholders as covered in Part II – 'Concepts and perceptions of value'
- the value chain models narrated in Chapter 4, under the heading 'Business value creation in digital economy'
- the need for directing data analytics investments to business strategy as in Chapter 6, under 'Data analytics (AI/ML) alignment with business strategy'
- value arising in supporting judgement and decision-making that we discussed in Part VII – 'Human vs automation with AI – An interdisciplinary framework', and
- the principles of measuring and attributing data analytics value covered in Chapter 7 under 'Data analytics (AI/ML) value attribution' along with 'Data leverage saves costs and speeds up time to value.'

We will address these with various methods, techniques and tools that include value assessment/realisation (DAVA/DAVR), value estimation (DAVE) and value optimisation (DAVO) covered in their respective chapters but we will first consider the principles and rationale behind these approaches, methods and tools.

Approaches and methods of data analytics (AI/ML) value measurements

There are a range of financial benefits that data analytics generates but these are not directly attributed and communicated within the organisation. As a result, they are invisible and hidden in the profit and loss statements (P&L), and in the organisation's balance sheet as described in Figure 38. The most visible part of data analytics is the total cost of ownership (TCO) of data analytics investments because these are easily quantified and monitored with a focus on reducing costs. On the other hand, the invisible and hidden aspects call for a range of methods and techniques to effectively measure, attribute and communicate the data analytics values which are addressed by the data analytics value framework (DAVF) in the next few sections and chapters along with the guiding principles and rationale behind them.

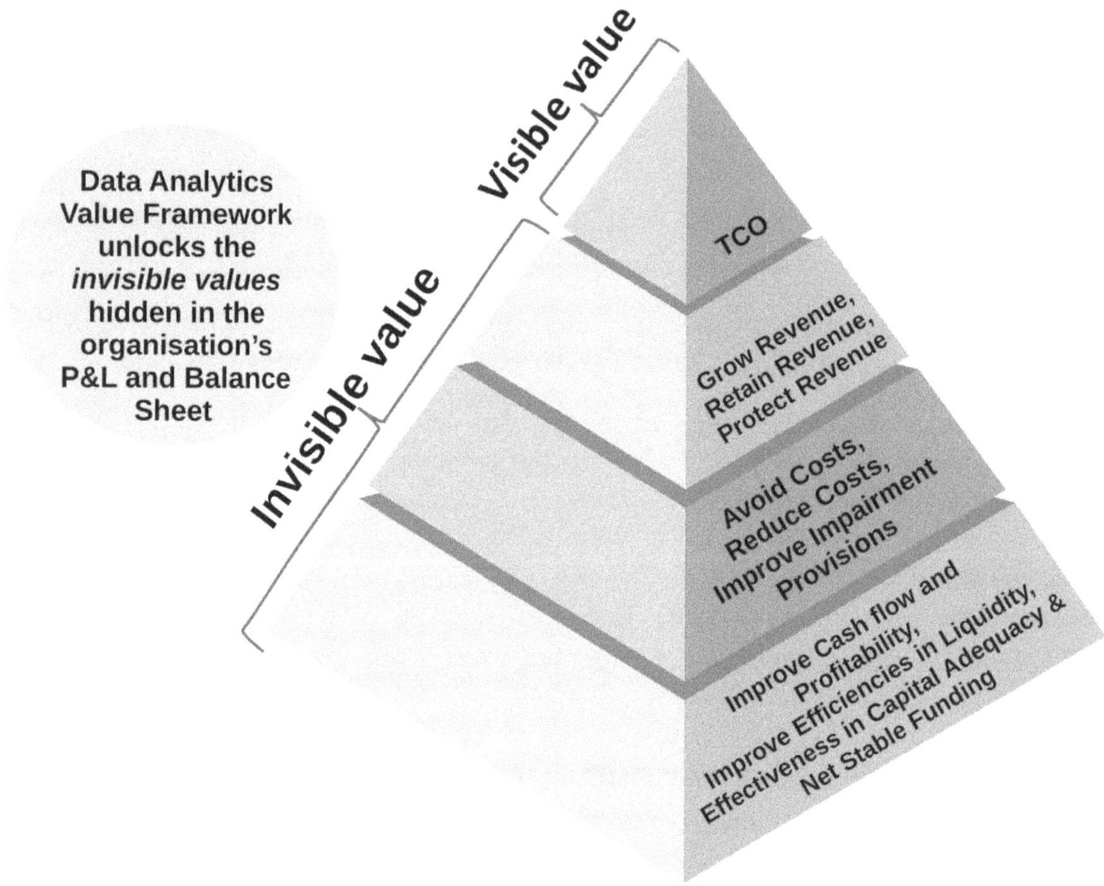

Figure 38: Data analytics value framework unlocks the invisible and hidden values

Top-down and bottom-up methods

To accommodate the different situations and perspectives prevalent in the organisation, the methods DAVA/DAVR, DAVE and DAVO mentioned earlier will employ top-down and bottom-up approaches to data analytics value measurement and attribution.

Tangible and notional value

The methods DAVA/DAVR, DAVE and DAVO will additionally use the guidelines in Tables 10, 11 and 12 in order to determine whichever of the 'tangible value' or 'notional value' is applicable for the data analytics value attribution method under consideration.

Table 10: Tangible and notional values of data analytics – 1 of 3

Tangible value	Notional value
How is this defined?	
Tangible value refers to the determination of data analytics value that can be linked directly to observable events as evidence for attributing its value.	Notional value refers to the probabilistic estimation of data analytics value that is linked indirectly to potential events that are not directly observable yet but, there is likelihood of their occurrence in a future. It is an opportunity cost that can be estimated as avoidance of the likely future event by using data analytics to prevent it.
Some examples with financial benefits enabled by data analytics	
Marketing campaigns for new customer acquisition (revenue growth), existing customer retention by offering incentives (retained revenue), cross-sell (revenue growth), up-sell (revenue growth). Data analytics value can be directly attributed from successful count of prospects and customers who took up the offers in the respective campaigns along with the proportion of incremental revenue generated by using data analytics that was instrumental in the campaigns.	Most regulatory compliance and reporting falls under this wherein the value of data analytics can be attributed to timely and efficiently delivering the required compliance timeline and frequency. This can be the case in many parts of the enterprise and includes business functions of: - Finance (e.g. Taxation), - Risk Management (e.g. probability of default), - Corporate Affairs (e.g. Law enforcement), - Technology (e.g. Network and platform resilience), - Customer Management (e.g. Privacy protection; Consumer rights compliance), - Marketing (e.g. Equality) etc.

Table 11: Tangible and notional values of data analytics – 2 of 3

Tangible value	Notional value
Methods of determining business value attributable to data analytics	
Because tangible value can be calculated from directly observable events these can be sourced from observable facts in the organisation's transactions, paper documents and digital event logs collected and retained by the organisation in their information systems and data analytics ecosystem.	Notional value can be calculated from observable events that have uncertainties associated with them. Therefore, it will require a combination of techniques to estimate the data analytics value from the observable events by applying probabilities. This includes probability estimate of potential non-compliance events or breaches, probability estimate of the costs incurred and/or revenue losses for the breaches, frequency of occurrences and the probability estimate of the likely future non-compliance and breaches. Historical trend for the industry from external sources along with history of similar events within the organisation as well as the type of non-compliance and breach will provide good indication for the probability estimates. Because these are estimates of probability they will require consultation with the organisation's stakeholders to seek their consent and mutual agreement of the data analytics value contribution that was instrumental in preventing such non-compliance.

Table 12: Tangible and notional values of data analytics – 3 of 3

Tangible value	Notional value
Other examples	
Revenue recovery from collections, avoidance of revenue leakage, revenue protection by detecting and preventing fraud. Data analytics value can be directly attributed as a proportion of revenue recovered, revenue leakage avoided, revenue protected etc by employing data analytics that was instrumental in revenue collections, revenue detection, fraud prevention etc. Total Cost of Ownership (TCO) improvements from moving on-prem data centre to on-demand Cloud computing resources.	Non-regulatory use cases where the value of data analytics can be attributed to predictions and/or probability of occurrences of events such as: - Finance (e.g. Financial impact of not closing the accounting books on time), - Insurance claims - Climate risk (i.e. flood, fire, drought) induced delinquency of mortgages - Predictive maintenance In general, non-availability or inability to access the required data analytics resources for extended period of time can be considered as an opportunity cost (e.g. platform failure/downtime; lack of required data sources).

How data analytics enables appreciation of business value

Unlike physical and material resources, data becomes more valuable when organisations refine it and combine with different data assets to extract new insights. Unlike machinery that wears out, when combining multiple ML algorithms, data analytics becomes more potent with improved predictions, and generation of new insights which help organisations to make better decisions that result in improved business performance. Figure 39 provides an illustration with an example from the financial services industry. Weather forecasts by weather bureau may have limited value for some but, when a bank combines weather bureau forecasts with its customers' mortgage loan data and physical location of the properties to determine credit risk of mortgage, it will be more valuable in taking actions that can prevent losses due to potential delinquency resulting from severe flooding. Together with advanced multi-genre analytics that can enable a data scientist to combine multiple algorithms to predict probability of flooding and probability of default, the bank can take proactive actions to offer new home insurance or top-up existing insurance to the homeowner. Natural language processing with integration of LLMs can help determine homeowner sentiment. At each stage in the data analytics step, the combined insights provide additional levels of business value by allowing the bank to stem its losses as well as helping its customers with financial protection. This secures the bank's revenue and provides better customer experience and retention.

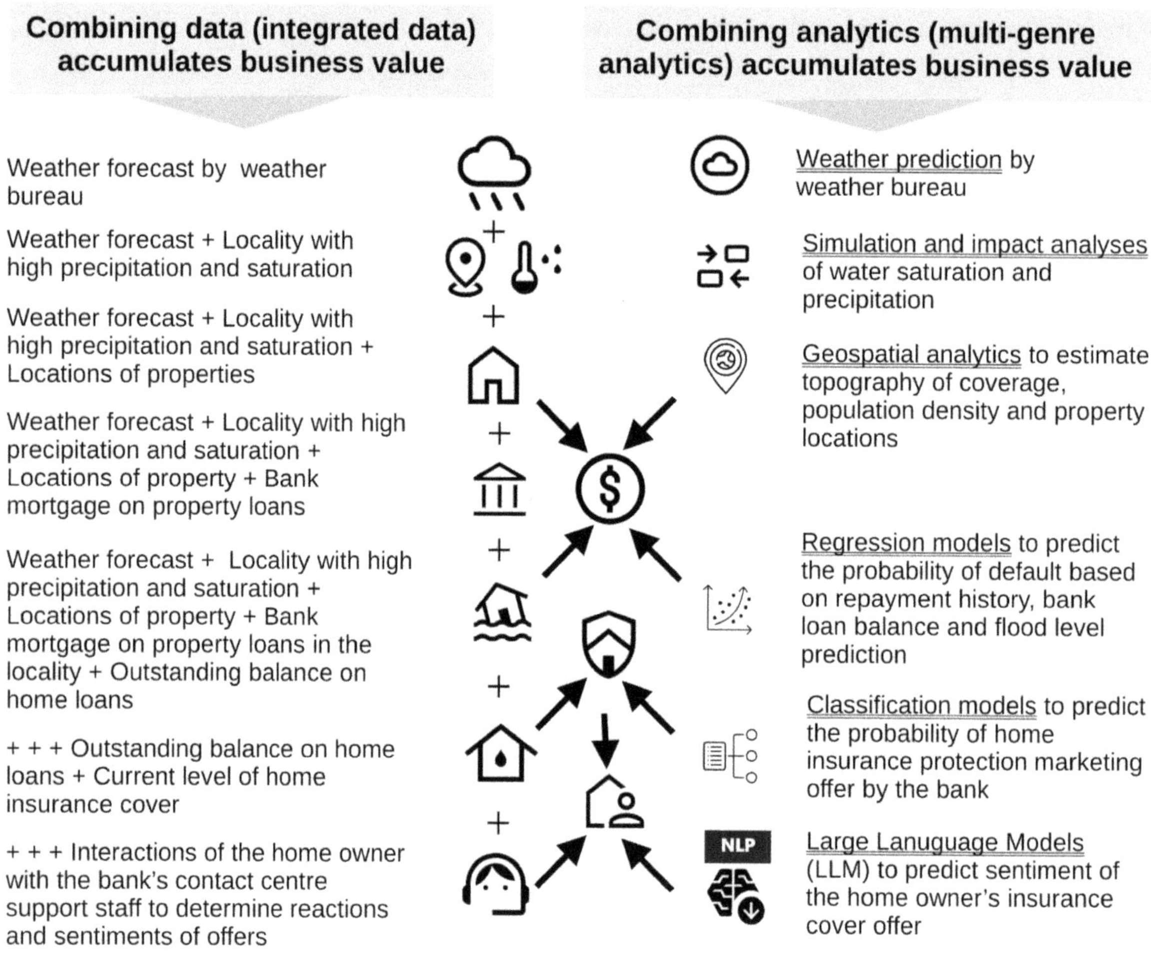

Figure 39: How data analytics enables appreciation of business value

Data analytics-driven KPIs induce cascading value to enterprise performance

As part of the planning process, each business function in the organisation sets their own KPIs that are aligned with the organisation's strategy and objectives, which the business unit managers monitor periodically to support decision-making. These KPIs are operational metrics that contribute to the organisation's financial performance, such as revenue growth, cost reduction, cost avoidance, and profitability.

Table 13 contains some example KPIs and their associated primary financial metrics, along with the business functions to which the KPIs apply. In capital intensive industries such as airline, manufacturing, telecom and utilities, the infrastructure and assets are their main product which these industries use to deliver services to their customers and citizens. In order to improve the KPIs and financial performance, the respective business functions initiate activities and tasks in the form of use cases to address their objectives with target metrics.

Table 13 provides three examples (circled with example number) of such undertakings that illustrate the cascading effect of data analytics value across the enterprise beyond the business function that initiates the use case. In example 1, 'Improve asset life time value' is focused on sweating the investments in the infrastructure by generating sufficient revenue from the asset to achieve profitability target. In this case, data analytics can be used to analyse capacity of the infrastructure, periodically monitor idle capacity and utilisation of the infrastructure and forecast future demand, thereby it can not only operate the infrastructure asset efficiently, but can also avoid additional infrastructure spend and grow revenue by creating demand through marketing efforts. In this example, the actions taken by the infrastructure business function help the enterprise with efficient infrastructure operation that results in profitability for finance business function.

Table 13: KPIs and associated financial metrics related to business functions

		Business Functions Management							
Key Performance Indicators (KPI)	**Primary Financial Metric**	**Customer & Marketing**	**Finance**	**Fraud & Security**	**Infrastructure & Asset**	**Legal & Risk**	**Operations**	**Product**	**Supply Chain**
Improve asset capacity and demand	Grow Revenue								
Improve asset life time value	Profitability				1				
Improve asset location and condition	Avoid costs								
Improve asset performance	Avoid costs								
Improve asset supplier quality	Reduce costs								
Improve asset utilisation	Grow Revenue								
Improve customer experience / NPS	Retained Revenue				2				
Improve equipment capacity/utilisation	Profitability								
Improve fleet utilisation	Avoid costs				3				
Improve reliability	Avoid costs								
Improve root cause identification of failures	Avoid costs								
Optimise asset maintenance	Reduce costs								
Reduce complaints	Reduce costs								
Reduce operational costs	Reduce costs								
Reduce parts inventory	Reduce costs								

Similarly, in example 2, the KPI 'Improve customer experience/NPS' may be the result of ongoing complaints from customers about not being able to connect their mobile phone to the telecom cell towers or frequent dropped calls that prevent them calling their friends and families. Customer complaint may also be related to dissatisfaction with the product or service (e.g. mobile phone or email service) or could be related to long waiting time before an agent answers the customer's call when attempting to interact with the contact centre about the dropped call or mobile phone problem. Operational KPI target metrics for these business units could be:

- improve Mean Time Between Failures (MTBF) from 500 hours to 650 hours or reduce unexpected faults from 6 to 4
- improve first call resolution by 15% or
- reduce call handling time from 180 seconds to 150 seconds.

Data analytics helps in analysing customer complaints to identify root causes of the complaint that can result in one or more initiatives for the respective business functions, who can take action to prevent the complaint. Predictive analytics helps to anticipate the problem and takes preventive actions that assist in not only reducing operational costs for the business unit but also improving customer satisfaction by preventing customer churn and retaining customer loyalty and revenue.

In example 3, the KPI 'Improve fleet utilisation' data analytics not only helps the infrastructure business unit but also operations management and supply chain management in reducing costs across the enterprise by optimising fleet and logistics operations.

Similar to Table 13, Table 14 contains example KPIs (circled with example number) that are common across many industries such as financial services, retail and telecom that drive flow-on effects of business value across the enterprise business functions.

Table 14: Cross-industry KPIs and associated financial metrics related to business functions

		Business Functions Management							
Key Performance Indicators (KPI)	Primary Financial Metric	Customer & Marketing	Finance	Fraud & Security	Infrastructure & Asset	Legal & Risk	Operations	Product	Supply Chain
Improve budgeting and forecasting	Profitability								
Improve campaign effectiveness	Profitability								
Improve Cash flow	Cash flow						4		
Improve Cash Operating Cycle (Days) - Purchase to Cash	Cash flow								
Improve channel effectiveness	Reduce costs								
Improve collections	Retained Revenue						5		
Improve contract compliance	Avoid costs								
Improve cross-sell revenue	Grow Revenue								
Improve customer sentiment / NPS	Retained Revenue								
Improve customer service / NPS	Profitability								
Improve Days In Inventory	Avoid costs								
Improve offer relevance	Grow Revenue								
Improve product bundling	Retained Revenue								
Improve sales revenue	Grow Revenue								
Improve staff competency	Grow Revenue								
Improve staff skills	Avoid costs								
Reduce channel costs	Reduce costs								
Reduce parts inventory	Reduce costs						6		
Reduce security threats	Reduce costs								
Reduce sensitive data usage	Reduce costs								

Communication of data analytics value

While much of the effort is devoted to measuring, quantifying and attributing the value of data analytics investments, it is more important to communicate and present the value contributions across the enterprise. This promotes the business capabilities that data analytics has enabled having made better informed decisions. There are other benefits and reasons for this communication, including to:

- demonstrate progress in aligning data analytics investments to strategic initiatives and objectives
- provide visibility and assurance to stakeholders that data analytics investments are providing measurable business outcome and performance that are not readily apparent in the profit and loss statement, and balance sheet
- help managers and executives increase investments in areas where there are greater returns and decrease or withdraw funding that is not effective
- advance data analytics capability levels through best practices in leveraging data, ML and AI technologies for effectiveness in enterprise performance.

Calculations of data analytics financial value

As discussed earlier in the chapter, key performance indicators are an effective measure of the benefits and value contributed by data analytics in improving operational performance of the business units. However, these KPI metrics will need to be associated with corresponding financial metrics to consistently measure if the investments are effective in their intended business outcome.

There is a range of financial measures that can be used to present and communicate to executives and stakeholders in the organisation. An effective way of communicating the financial benefits of data analytics is to present it in the form of ROI using discounted cash flows techniques, and proforma profit and loss of data analytics which are shown in Figure 40.

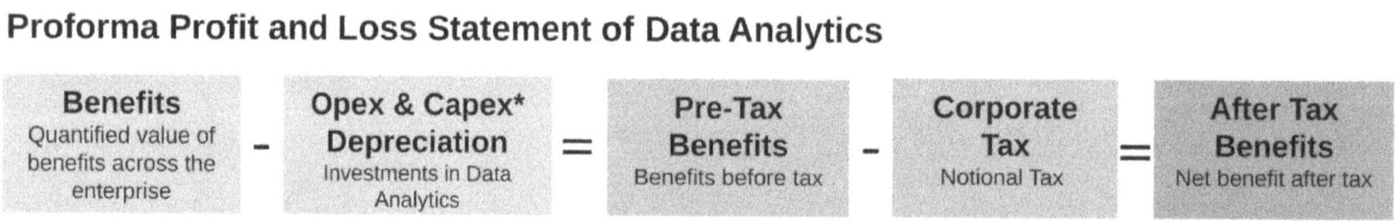

Capital expenses (Capex) are shown as **depreciation expenses per year as per straight line depreciation rate of nn%. **Corporate tax** is the notional tax payable by the organisation as if the funds were invested in generating revenue*

Net Present Value (NPV) is calculated by applying **corporate hurdle rate of nn%** to incremental net cash flows. When NPV is greater than 0 then the investment is desirable.

***Net Present Value** represents the net notional value that the organisation would gain after deducting taxes from net benefits and after adjusting for cost of borrowing, investment and inflation risks as if the funds were invested in generating revenue.

Internal Rate of Return (IRR) is also calculated from the incremental net cash flows. IRR of greater than cost of capital (i.e. Hurdle Rate) then the investment is desirable.

Figure 40: Calculations of data analytics financial value

Proforma profit and loss of data analytics is presented as financial benefits accrued per year minus capital expenditure (Capex) which is the depreciation expense for the year minus operating expenses (Opex) incurred in that year, the sum of which gives pre-tax benefits. In Figure 40, depreciation expenses are shown using straight line depreciation rate of nn%

but, any depreciation/amortisation method as per the organisation's policies may be applied. Corporate tax is deducted from the pre-tax benefits to present the after-tax benefits. Corporate tax is the notional tax payable by the organisation as if the funds were invested in generating revenue.

Proforma discounted cash flow of data analytics presents the flow of funds and working capital to support the investments in data analytics by taking into account the time value of money. It calculates the net present value (NPV) after considering financial benefits as inflows, capital expenditures, operating expenditures, and corporate taxes as outflows to derive NPV. NPV represents the net notional value that the organisation would gain after deducting taxes from net benefits and after adjusting for cost of borrowing, investment and inflation risks as if the funds were invested in generating revenue. NPV is calculated by applying the corporate hurdle rate of nn% to incremental net cash flows. When NPV is greater than 0, then the investment is desirable.

Internal Rate of Return (IRR) is also calculated from the incremental net cash flows. If the IRR of greater than the cost of capital (i.e. Hurdle Rate), then the investment is desirable.

The proforma profit and loss and the proforma discounted cash flow reports are presented on a rolling seven-year basis (i.e. most recent seven continuous years) that include current year together with three years of historical and three years of future cost and value projections. While it is desirable to present data analytics value for the most recent seven years, some organisations may only be interested in current and future years or unable to quantify historical values. In these cases, the three years of history in optional. The actual process and method of quantifying the cost and value will be covered in detail in their respective chapters for DAVA/ DAVR, DAVE and DAVO methods mentioned previously.

Presentation of data analytics financial value

One of the primary premises of this book is that data analytics continues to generate recurring and long-term value which is not apparent in the organisation's financial reports such as the balance sheet, where data analytics does not appear as an asset due to the constraints imposed by the accounting standards (i.e. IFRS and GAAP). Throughout the book many chapters discuss and illustrate how data analytics contributes to the long-term and cumulative financial benefits to the enterprise. Among the alternative mechanisms to present the long-term value of data analytics, the chart in Figure 41 has been found to be effective at all levels in the organisation during the many customer engagements, to communicate and present the data analytics value.

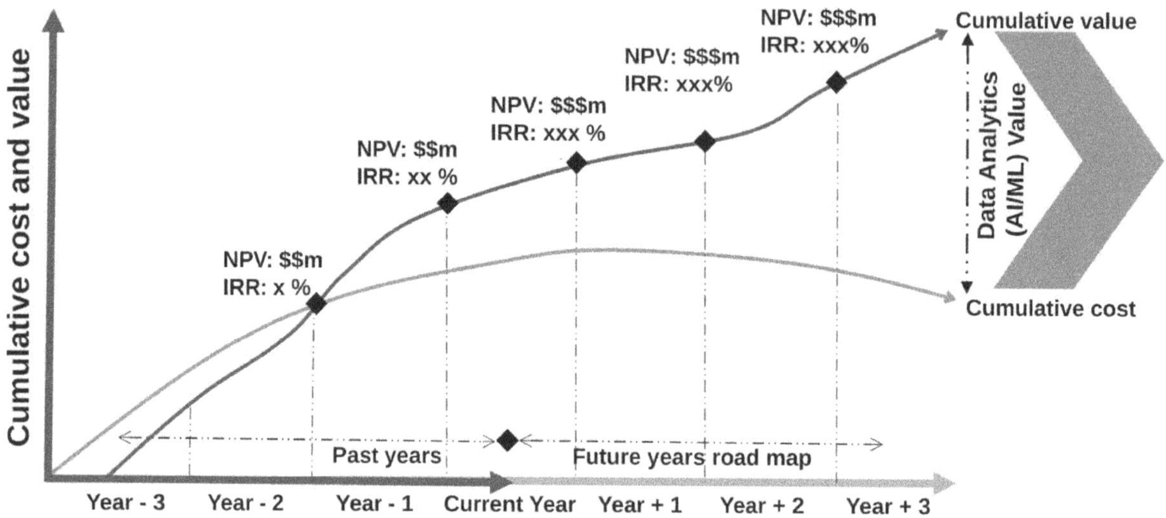

Figure 41: Strategic long-term value enabled by data analytics (AI/ML) [Adapted from © Teradata, 'Value of Integrated Data']

Figure 41 presents the data analytics value as cumulative costs and cumulative value using the financial calculations shown in Figure 40. However, one of the key differences between costs and values is that the business value compounds exponentially, due to factors we discussed in the previous sections, such as data reuse and the cascading effect of KPIs, whereas the investment grows only incrementally thereby providing compelling ROI. The costs and value are presented on a rolling seven-year basis (i.e. most recent seven continuous years) that include current year together with three years of historical and three years of future cost and value projection. As mentioned earlier, three years of history is optional as some organisations may be only interested in or capable of current and future year values. The NPV and IRR values for the respective years are enterprise level summary extracted from the proforma profit and loss, and the proforma discounted cash flow reports that contain the details of the various data analytics initiatives delivered and/or in the road map. Examples of the detailed proforma profit and loss, and the proforma discounted cash flow reports are covered in the respective chapters for DAVA/DAVR and DAVE methods.

One of the rationales for choosing the seven-year rolling cycle is that while data analytics continually contributes to recurring value lasting for many years, the corporate memory of managers and executives who are responsible for it fades due to dynamic and periodic changes in organisational and business restructures that result in corresponding changes in their scope, motivation and interests. Besides this, the knowledge base of people who were involved in the past data analytics initiatives disappears, resulting in difficulties finding suitable personnel to acknowledge and confirm the benefits that can be confidently presented to stakeholders and executives. To put this in perspective, here is what one of the data centre operators said when he was asked to explain the reasons for specific analytics among others that he runs, the business benefits and the beneficiaries. He said, '*I run these analytics every night because when I took over the role, this analytics job was in the list of scheduled jobs. No one was able to explain why this was required!*' There were other meetings in which the responsible staff were unable to explain the dashboards and reports they produce.

The seven-year cycle has been effective as it provides a reasonable level of stability in the organisation to be realistic in finding responsible persons with good knowledge of the objectives and benefits of data analytics initiatives.

CHAPTER 17

Data analytics value assessment

Overview: The focus of this chapter is to describe the process and method along with examples of data analytics value realisation which is also known as value assessment or value audit.

Data Analytics Value Assessment/Realisation (DAVA/DAVR)

Data analytics value assessment or value realisation (also known as the value audit) is part of the data analytics value framework. It is a top-down approach to quantify value contributions of data analytics investments that have been in deployment for a period of time. It establishes a direct link between data analytics, business outcome, and its value contributions. Its emphasis is on data analytics-driven business decisions and actions by actively working with stakeholders, managers and executives to ascertain their perspective and perceptions of data analytics initiatives, expected outcome, and delivered results. It is also a method used for discovering new initiatives, which is useful in developing a road map of future values of these initiatives. Due to the extended time and efforts needed for this work, the frequency of the data analytics value assessment can be once a year or once every two or three years. More frequent short assessments or evaluations will be required which is covered in Chapter 20 under 'Governance oversight'.

Besides being part of best practices and governance, there are a number of other reasons why data analytics value assessment is important. These include situations such as:

- a perception within business units who fund data analytics projects that these investments are not seen to be delivering value
- visibility of value from data analytics investments at senior executive levels is limited or absent
- organisational restructure or change in leadership whose primary interests may be optimisation of costs that often leads to questionable value of data analytics investments
- business units that struggle for justifying investments in new initiatives and
- some business managers are genuinely interested in quantifying the value, but do not have the necessary skills.

An ideal way to perform data analytics value assessment is to engage with the users, managers and executives who are or have participated in one or more of the activities involving data analytics initiatives and/or investments, and/or in their use and/or in their operation. This is best done by means of interactive meetings which allow the opportunity to identify their motivations, the process of how data analytics is being used for decision-making, as well as to spot bottlenecks that may be inhibiting value. In order for the meeting to be effective, prior planning, and research of the data analytics operational environment (i.e. usage of platform and software resources), analytics development process, and its usage in business applications will be required. Meetings with members of business units and technology units will need be scheduled for the assessment to be fruitful.

A high-level process of data analytics value assessment is shown in Figure 42 that outlines the activities involved in the assessment. It encompasses a review of data analytics solutions, processes, KPIs and their application in the respective business functional areas. It also comprises quantification of tangible and notional values of data analytics by isolating from the effect of other factors (i.e. excluding value contributions of people and other resources in the organisation – see Figure 16 and Figure 17) when quantifying the value.

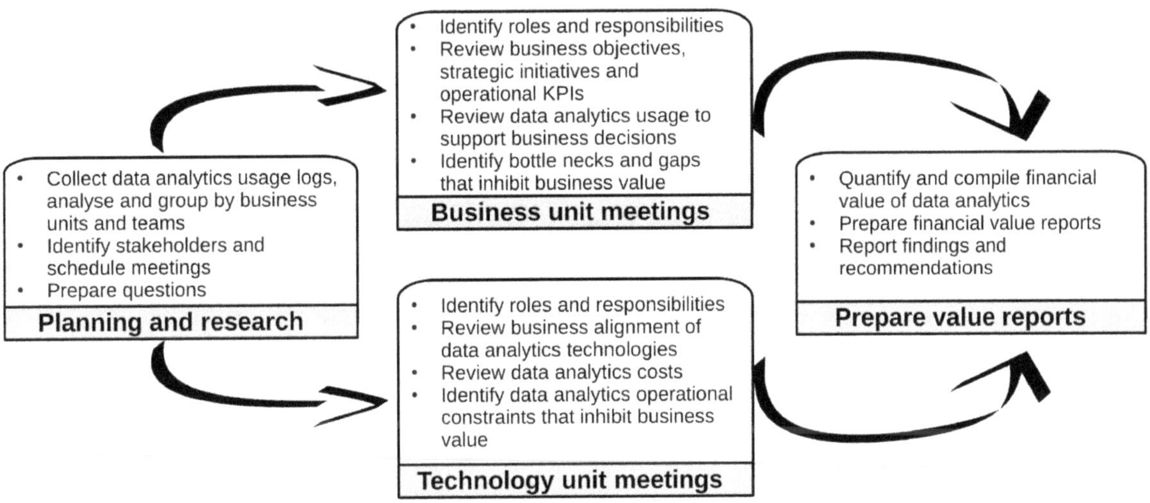

Figure 42: Data analytics value assessment process

Planning and research

While meetings are an effective way to interact with members of the business and technology units and solicit information, they are not a substitute for research and analyses of the data analytics ecosystem and analyses of their application in organisational decision-making. Periodic snapshots of data analytics usage in data centre operations and usage of data analytics applications across the enterprise business functions will provide a holistic and cumulative view on which value assessment can be based. This also avoids the shortcomings of the approaches used for transactional systems. By comparing periodic snapshots it also provides a view of usage over time which helps to identify those resources and applications, whether they are in active usage, on the verge of retirement, or not in use. This will allow optimisation of the resources through improvement opportunities. The meetings with the business units and technology units can be used to get their confirmation of those data analytics resources that are live and effective in their day-to-day operation, as well as those that are not in continued use in the organisation.

While it is acknowledged that the data analytics ecosystem is complex, requiring a range of usage statistics to be collected and reported, Table 15 and Table 16 provide examples of the aggregated view of resource usage that can be useful in dialogues with members of the business and technology units to confirm and take measures to optimise resources. They also provide an effective way to measure the business benefits and value to the organisation at user group and business sub-function levels.

Although higher usage of CPU, memory or disk storage as in Table 15 are not an indication of higher value, they provide useful information in correlating with their effectiveness in delivering business outcome. It also provides an opportunity to optimise these resources, if they are due to inefficiencies in the analytic development and/or operation.

Table 15: Data lakehouse resource usage

Weekly Average Data Analytics Usage by User Group (over 20 week period)			
User Group	**Average Disk Usage (GB)**	**Average CPU Usage (Secs)**	**Average Memory Usage (MB)**
Analytics (AI/ML) Tools and Applications	27,427	38,830	10,378
Billing and Accounting	5,713	4,283	1,009
Business Intelligence	4,426	7,181	1,683
Customer Operations	13,719	14,070	3,112
Data Management (ETL)	11,099	12,379	4,397
Data Platform Resources	40,518	39,948	9,187
Discovery Analytics	18,356	18,442	5,769
Distribution and Logistics	1,799	1,434	264
Enterprise Reporting and Dashboards	64,539	40,148	8,549
Finance	2,407	1,612	419
Human Resource	629	3,423	4,144
Infrastructure	3,667	5,020	1,791
Marketing	4,384	6,747	773
Pricing	1,675	4,491	1,616
Product Management	1,356	4,619	1,294
Regulatory Compliance	2,002	3,711	753
Sales	1,609	1,872	333
Security and Risk	667	1,861	3,872
Statistics Collection	7,291	17,198	3,705

Table 16 provides a good indication of how data analytics is being used in business applications that generate business outcome and value. The list allows the opportunity to discuss with members of the business units to ascertain how these relate to enterprise business strategy and objectives.

Table 16: Data analytics application usage

Data Analytics Application	Average Weekly Usage (No. times)	Data Analytics Application	Average Weekly Usage (No. times)
Abandoned Online Purchase	103	Executive Information Hub	178
Accounting Operations and Close	126	Field Operations	69
Advertising Creative Optimisation	40	Financial Reporting	10
Asset Failure and Maintenance	158	Improving Cross-Sell Targeting	234
Asset Impact Analysis	106	Insight Driven Customer Management	226
Assortment Range and Space	118	Inventory Management	242
Call Centre Sentiment	101	Lead Generation From Call Centre Notes	167
Cash Flow Optimisation	73	Operational Change Management	42
Complaint Analysis	187	Planning and Demand Management	47
Credit Risk Models for Lending Customers	40	Pricing and Promotion	166
Cross-Channel Customer Engagement	105	Procurement and Sourcing	27
Customer Interaction Management	183	Product Development	49
Customer Journey Analytics	132	Supply Chain Finance	81
Customer Life cycle Management	25	Supply Chain Network Management	164
Digital Marketing Investment	152	Valuation of Intangible Assets	200
Enterprise Performance Management	153	Workforce Management	55

Meetings with business and technology units

The main purpose of meeting with business and technology users, and managers is to confirm business goals/objectives, and identify tangible improvements. These improvements include revenue growth, profitability, cost savings, cost avoidance, productivity, and customer satisfaction that are clearly tied to the organisation's use of data analytics, and the resulting business value.

The data analytics value assessment is conducted by meeting with the users, managers and stakeholders throughout each of the enterprise's business groups/functions to ascertain the use and value derived by each business group/function. The meetings with executives and senior managers will be focused on exploring their organisational strategy and goals, and their perception of the data analytics technology and tools as value-enablers. For business users and analysts, the meetings will be used to confirm and document quantification of benefits achieved from decisions and actions taken by using data analytics resources.

While quantifying the benefits, the process will allow the isolation of other factors that contribute to the business unit's functional effectiveness and focuses on information-enabled/analytics-enabled values contributed by data analytics resources. Where tangible and quantified value is difficult to determine, the meeting will seek to identify notional value of business benefits realised by the business units.

Meetings with members of the technology unit will provide opportunities to discuss how IT supports business alignment. The meetings can be used to assess costs of data analytics resources, as well as any inefficiencies and bottlenecks that prevent technology teams from delivering results and business outcome.

The information documented during the meetings is then collated and presented back to each area with the findings and assumptions made. Documentation from the meetings and quantified value reports prepared following this will allow business managers to establish a baseline of how their funding for investments in data analytics contributes to business outcome which is generally invisible and hidden in the balance sheet of the organisation. The quantified value reports will enable clear visibility of the data analytics value generation process and will provide a methodology to prioritise investments. It is also expected to simplify future business cases to be considered in the light of past ROI.

Standard questionnaire and participant responses

In order for the meetings to be effective and to ensure consistency in information gathered during the meetings, a standard questionnaire will be useful. A template for the questionnaire, as well as samples of responses from the participants from a very large financial services organisation, have been included in Tables 18 to 22 that contain varying degrees of participants' perceptions of value, data analytics success, bottlenecks and so on, which we will discuss in detail in the next few sections of this chapter.

The questionnaire in Table 17 is used as a prompt during the meeting to cover various subjects that include: Roles, key tasks, and processes used by the business unit users and managers to make decisions against core objectives by leveraging data analytics; exploration of value drivers and key performance measures; identification of bottlenecks that prevent value being achieved as well as future requirements for a road map.

Table 17: Template for standard questionnaire

Categories of questions	Meeting questions
Key Tasks / Processes / Decision and Action	What are the key tasks and processes you / your team perform? How is Data Analytics helping you to make decisions and take actions? Do you distribute the results of your analysis? If so, how are the recipients benefiting from the analyses?
Business Value Drivers	What amount of your function's/project's effectiveness depends on Data Analytics and how do you quantify the value attributable to Data Analytics? What would you have done/used if there were no Data Analytics in your business area? What is the alternative solution? What other factors contribute to / constrain your function's overall effectiveness?
Operating Objectives / Metrics / KPIs	What are your operating objectives and KPIs? What % contribution from data analytics is enabling you to achieve your KPIs?
Future Data / Analytics Requirements	What would you like to do that you are not able to do at the moment and what are the bottlenecks? What is your vision / requirements for the use of next three years?

Table 18: Questions and responses from Digital Marketing business function

Categories of questions	Participants' responses
Key Tasks / Processes / Decision and Action	Role: Digital Marketing Targeted and relevant message presented to customers via digital channels Core Objectives: To monetise the digital interactions by: - Up-selling; On-boarding; Funding and effective use of products - Customer retention; Lower service costs using digital channels
Business Value Drivers	"No data analytics platform would mean there is no digital marketing program" 200 active campaigns each one comprises: - Who to target - inclusions and exclusions - Trigger - Propensity models (score), Event, Transaction - 30 m impressions a week
Operating Objectives / Metrics / KPIs:	KPIs: 1. Ultimate KPI is to increase Operating Income 2. Increase customer deposit balances 3. Reduce servicing costs Operational KPIs include: - Post click metrics; Does someone click on the presentation - % click to population; % click to sales - How many times presented (passive presentation) Key Value Measure: Consumer Operating Income The value of data analytics attributable to the digital marketing as central information asset is 1.5% of Digital Marketing sales success which translates to $30 million per annum.
Future Data / Analytics Requirements	Issues: Platform performance is slow; reliability and timeliness is below standard - All of this affects general efficiencies of staff. Need real time loading and decisioning to move from the current segment level marketing to the desired hyper-personalisation. Real time will enable us to provide an uplift from the current 1.5% Digital Marketing sales success to 5% which translates to an additional $70m per annum.

While some participants tend to have a clear idea of how to articulate the value contributions (see Table 18), others may need help in doing so. The meeting can be used to educate the participant about a method they can use to help make progress.

For instance, Table 19 contains an example from customer management business function, and the meeting participants found it difficult to determine how to quantify value of customer retention, cross-sell activities, and customer experience measurement. In these cases, industry benchmarks will assist in providing an estimate of the value, which we will cover in Chapter 18. The same table also provide examples of other value inhibitors, such as poor data quality, and lack of standardisation that affects value.

Table 19: Questions and responses from Customer Management business function

Categories of questions	Participants' responses
Key Tasks / Processes / Decision and Action	Role: Customer Management Team Size: 20 (approx) Team Objectives: - Understanding the customer; Identifying customer needs; - Data operation and advanced analytics (modelling, segmentation, etc); - Campaign execution and effectiveness; new technology and automation Focus is customer prospecting and retention.
Business Value Drivers	We can quantify the value as: The absence of the data analytics would mean we would need to go to myriads of data source systems to collect the data we need. Estimated impact is we would need 5 to 10 times the staff to accomplish the same outcomes. **Efficiency Value = 90 FTEs @ $125K p.a. (fully costed) = $11.25m per annum**
Operating Objectives / Metrics / KPIs:	The primary measure impacted by our team is Net Operating Income. We impact this in a number of ways: **1. Campaign enables sales (prospecting measure)** A value measure could be computed by the following equation: Enabled Sales Count x Return on Marketing Investments = Value **2. Retention and up-sell to existing customers** A little more difficult to quantify. **3. Customer experience** Quite difficult to quantify. "Campaign effectiveness measure has been poor as we are in growth phase"
Future Data / Analytics Requirements	Key issues that create inefficiency and hinder usage of data analytics and processes include: 1. Customer Mastering - key attributes of data required for exploratory analyses missing 2. Contact History (located in multiple places) 3. Availability and timeliness of data 4. Usability and cost - Takes a long time to get new requirements into production and at high cost. We are not agile! 5. Lack of standardisation of tools and processes 6. Data analytics ecosystem is "Finance-centric" and needs more focus on the customer and their interactions Qualitative flow-on benefits include: 1. Intelligent conversation with customers will lift relevance of communication. 2. Positive impact on conversion rate.

Other participants (see Table 20 and Table 22) echo similar concerns as those in Table 19 about lack of data analytics governance, high costs involved in data acquisition, and so on. This is in line with the prerequisites for data analytics foundation that we discussed in Chapter 10. As to how we can optimise costs of data acquisition and management, data leverage and reuse are an effective mechanism to achieve this goal which we will deal with in greater detail as part of Chapter 19.

Table 20: Questions and responses from Consumer Analytics function

Categories of questions	Participants' responses
Key Tasks / Processes / Decision and Action	Team Role: Consumer Analytics Team Size: Onshore 54; Offshore 16 BAU Role: Support for - Marketing Campaigns with retention and cross sell; Customer profitability; Channel optimisation; - Product quality; Product usage - Propensity modelling for prediction, segmentation, self-service dashboard - Advisory services including industry and business analyses, market research and recommendations
Business Value Drivers	"We only exist because of data". "It is impossible to do our work without a centralised data analytics ecosystem". "We see a lot more potential". Delivering the same capability without a centralised data solution would involve 10 times the current FTEs (Full-time Equivalent). Efficiency of data analytics value can be estimated as: Onshore = $72.9M Offshore = $7.2M Total = $80.1M
Operating Objectives / Metrics / KPIs:	KPIs: Cross-sell and retention - proxy for share of wallet; - Net customer growth (depth of wallet measure); - NPS engagement; Channel migration (cost reduction) Effectiveness of Data Analytics Valuation Assessment BAU role: 1% of Consumer Operating Revenue can be attributed to data analytics with value calculated as $8.021B x 1% = $80.21M per annum Project role: $30m per annum of identified benefits
Future Data / Analytics Requirements	Key challenges (time wasters): - Inconsistencies in definitions and labelling of data (e.g Definition of a new customer) - Separate data platforms - New data is hard to acquire, very expensive and time consuming - Ownership of data is an issue Overall, if these issues were addressed, 40% of the team work would be saved with potential value assessment of Onshore = $3.24M; Offshore = $0.32M; Total = $3.56M

Table 21 contains responses from the technology unit team. This team supports the business units with information design, develop analytics, and delivers reports. The responses contain an organisational problem that we previously covered highlighting the disconnect between IT and business units, wherein the technology teams do not have visibility of how the business units use the reports to make decisions. The meeting notes also highlight the fact that, due to lack of visibility of how the business functions use the reports, there is potential for wastage of resources that are not leveraged well, as noted by the participant's comment, *'There is too much information!'* One of the purposes of the data analytics value assessment is to uncover such data analytics value inhibitors, so actions can be taken to optimise investments.

Table 21: Questions and responses from Business Analytics/Reporting support

Categories of questions	Participants' responses
Key Tasks / Processes / Decision and Action	Role: Responsible for Information Design, Analytics and Reporting support to the businesses Key tasks consists of: Analyses and advice on trends and forecasting of everything Reporting including executive reports and dashboards A team of 45 onshore and 10 offshore Over 700 reports are produced covering Financial (Front book; Back book) Non-financial (Depth of Customer Relationship; Customer Satisfaction/NPS)
Business Value Drivers	"Without this central asset we would require 15-20 additional onshore analysts." Data Analytics Efficiency Value Calculation: 8 Snr Analysts @ ($90k - $120k + 28% on costs) = $1.08m per annum. Plus 8 Analysts @ ($80k - $100k + 28% on costs) = $0.92m per annum Total $2.0m per annum.
Operating Objectives / Metrics / KPIs:	KPIs (Dimensions and Measures) are defined by the Business. We provide end to end reporting for the following business units: Consumer Bank; Business Bank; Operations; Distribution; Product Marketing / Segmentation. How the business units are using the reports to take action and make decisions from the insights from reports are not visible to us.
Future Data / Analytics Requirements	"There is a big problem. Lots of insights but no action". We produce a lot of reports. Do not know why the reporting is done. This year we started the charge back to the business units which we have not had previously. Future vision: There is too much information. We need to develop "Metric Map". Rather than reporting we need to focus on what is important and story board the information with the Business deciding what is important. Cost optimisation: 1. Improvement in performance would reduce costs by (10-15%) = $0.561M per annum 2. Breadth of Skills required = $0.25m per annum

Notional value of data analytics serves a dual advantage

Among the challenges in quantification of the value of data analytics investments, lack of visibility in the organisation's balance sheet, and shortcomings in financial reporting standards dominate as a key obstacle which we previously identified. As a consequence, data analytics is often considered as an overhead cost that attracts constant scrutiny with a focus on cost reduction.

In other situations, it is considered as essential business as usual (BAU) expenditure with its orientation as cost of doing business, such as in meeting regulatory compliance requirements. In such cases we recommended using the notional value of data analytics investments as explained in 'Tangible and notional value' (in Chapter 16), to provide visibility to the organisation. We will discuss these next.

Table 22: Questions and responses from Pricing business function

Categories of questions	Participants' responses
Key Tasks / Processes / Decision and Action	Team Role: Pricing for consumer products Team Size: 8 Key focus areas: - Analytics, Reporting & Insights; Historical monthly facts for 5 year history; Rolling weekly reporting for 17 weeks - Quantitative customer behaviour in relation to pricing; customer sensitivity/insensitivity to pricing changes - Customer needs to determine where we should / should not compete - Segmentation of granular data by a large number of possible pricing attributes - Pricing models to predict profitability of portfolio based on pricing assumptions such as margin, attrition, customer reactions Distribution is done by automating the publication to support self-service.
Business Value Drivers	"Could be left behind the market". "How fast we can deliver insights is important". We would need 32 analysts instead of the 8 people to do our job if the data analytics environment did not exist. Analytics Efficiency Value = (32 - 8) * $150K = $3.6M per annum
Operating Objectives / Metrics / KPIs:	Primary KPI: Consumer Operating Income Other KPI's: Balance sheet value (deposits and liabilities) Analytics Effectiveness Value is calculated as follows: Scope of Pricing Impact: 10% growth in Revenue Target Impact of Pricing on revenue growth: 60% Impact of data availability on pricing function: 30% Value = Consumer Revenue x 10% x 60% x 30% = $8.021b x 10% x 60% x 30% = $144m per annum
Future Data / Analytics Requirements	The following are inhibiting factors affecting value: - Breadth and depth of data – missing transactional level data - Timeliness of data - No "single source of truth" that results in additional cleansing and reconciliation activities at our end What improvement I would like to see: - Governance including controls for the following: - Removal of data duplication - Avoid data silos and create single data ecosystem - Do and don't principles of data model design - Business data glossary of metadata No potential future value identified

In all of the responses in Tables 18 to 22, the question, '*What amount of your function's/project's effectiveness depends on data analytics and how do you quantify the value attributable to data analytics?*', acts as a dual advantage in not only determining the value contributions of data analytics investments, but also in answering the key dilemma on which much of the premise of this book rests, namely, *Invest in human or artificial intelligence?*

The responses from the participants provide some clear answers:

- '*No data analytics platform would mean there is no digital marketing program*'
- '*How fast we can deliver insights is important*'
- '*We only exist because of data*'
- '*It is impossible to do our work without a centralised data analytics ecosystem*'
- '*We see a lot more potential*'
- '*The absence of the data analytics would mean, we would need to go to myriads of data source systems to collect the data we need. Estimated impact is we would need 5 to 10 times the staff to accomplish the same outcome*'

The summary of responses from the participants highlights the fact that:

- Data analytics is an essential resource, much like other technologies in the enterprise (e.g. email, knowledge sharing) and collaboration technologies (e.g. SharePoint, GitHub). Absence of data analytics resources and/or its unavailability for an extended period of time (e.g. system down-time) hurt the business units in achieving their tasks, objectives and the organisational outcomes
- Data analytics resources are under-utilised, and there is huge potential for its further use in applications in the business. Furthermore, assurance of data quality, and data analytics governance are prerequisites for driving more value
- Similar to other technology resources, data analytics has substituted many routine tasks traditionally performed by humans (e.g. collecting and compiling large volumes of data from several sources) by enabling cost efficiencies, and productivity in the enterprise and in acting as a competitive differentiator. Data analytics, therefore, contributes to efficiency value, which is represented as the notional value in many cases, but also quantified as tangible value, where it can be directly attributed

- Also, very similar to other technology resources in the enterprise, data analytics augments human tasks to drive effectiveness in decision-making, thereby delivering business outcome and value
- Data analytics (AI/ML) is here to stay in the organisation for a long time to come, but needs to be well-governed in order to continue to be a valuable resource to the organisation
- The ratio of data analytics in the resources mix depends on what percentage of the business function's effectiveness relies on it to deliver business outcome. In jobs involving routine tasks, we can expect a higher ratio of data analytics in substituting humans with Agentic AI-driven robotic process automation. In most other tasks involving inductive reasoning and expert thinking, data analytics is expected to augment human tasks in varying proportions as shown in Figure 43, subject to assessment, and expectations of risks, and rewards in business outcome. Human in the loop (HITL) is preferred for tasks involving higher risks and/or higher rewards.

Consider an example from Customer Management business function. The effectiveness of the customer management team in achieving the revenue growth objectives is partly determined by how well the team designs campaigns and develops go-to-market strategies. It is also fortified by segmentation and targeting accuracy reflected by the 360° view of the customer in the data analytics ecosystem and the prediction models employed about customers' propensity to take up campaign offers. Additionally, other technology resources, viz. email servers, social media, and the Internet, are also used to augment the launching of campaigns. The mix of resources used in the right proportion enable human Customer Services Representatives and/or Agentic AI to take next best action for the customer, in real time, to maximise customer experience, sales and customer retention.

In short, the answer to the question, '*Invest in human or artificial intelligence?*', will be not either this or that, but rather in establishing what intelligence is really worth and allocating investment accordingly by considering the risks and rewards.

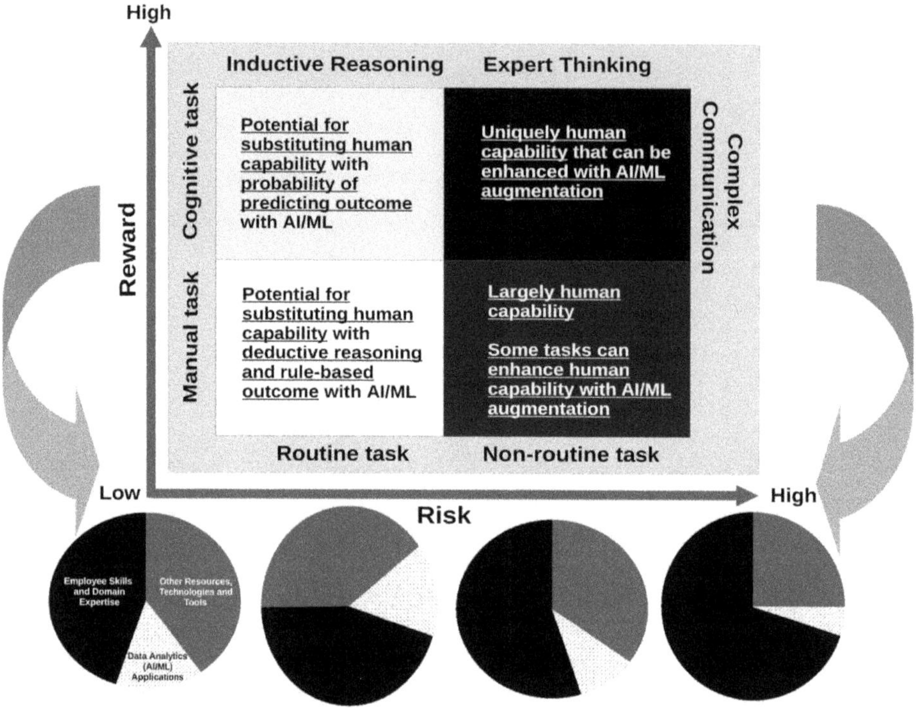

Figure 43: Ratio of data analytics use and cognitive capabilities required in job tasks

Preparing and communicating data analytics value reports

The primary purpose of preparing, and communicating data analytics value is to create awareness, and visibility of the value being driven from using data analytics throughout the organisation to deliver business outcome. The business outcome includes better understanding of customer journey, and experience, improving product profitability, and streamlining operational processes. Besides value reports, it is also important to document, and communicate blockades in the business, and operational processes that prevent or slow down value creation. This should be accompanied by recommended courses of action.

Some examples of the reports presented to C-level executives, and managers of a large telecommunication service provider in the Asia Pacific region is included in Figure 44 and Tables 23 to 26, which we will discuss in detail next.

Figure 44 provides a high-level view of the long-term value that data analytics investment generates which is invisible in the balance sheet of the organisation. As mentioned previously this chart is presented on a seven-year rolling cycle as ROI with cumulative value and costs. The chart also contains key data analytics initiatives that were undertaken each year that generate durable value to the organisation, and it accumulates during each subsequent year due to enduring and repeatable usage. By aggregating the value each year, the presentation also offsets against any data analytics usage that may be discontinued while also including new initiatives that expand usage and value. This also includes estimates of future value based on data analytics initiatives that are in the road map that are aligned with the organisation's strategic goals. It provides visibility to executives to support justification for further investments. A notable aspect of this presentation is that the business value grows exponentially (i.e. due to the unique characteristics of data) while the costs increase/decrease incrementally contributing to greater net value. Cumulative costs can decline over time if the organisations govern data assets well, including data reuse, which will be the main topic in Chapter 19.

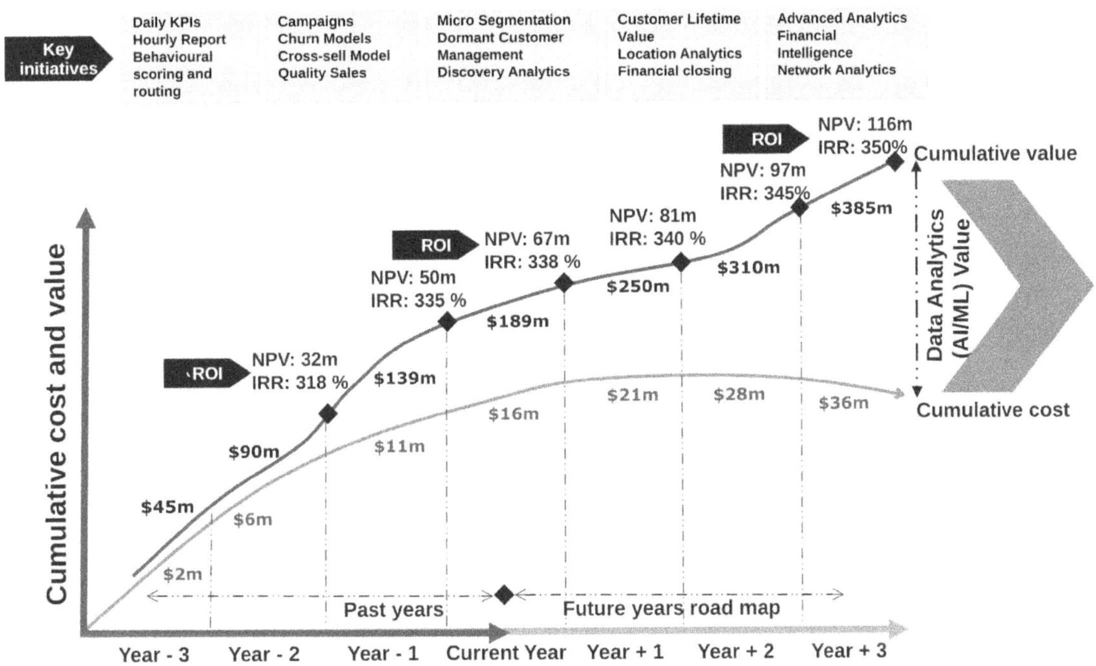

Figure 44: Strategic long-term value enabled by data analytics (AI/ML) [Adapted from © Teradata, 'Value of Integrated Data']

Table 23 represents consolidation of the business value and costs that have been discussed and confirmed from the meetings of business units and technology unit representatives. It is presented as proforma profit and loss statements including notional corporate tax to demonstrate that the investments in data analytics are delivering acceptable ROI compared with potential alternative avenues of investments elsewhere in the organisation to generate revenue. Examples include, opening a retail store or developing a new dealership channel for the organisation's products and services. The statement breaks down data analytics value by divisions, and business functions where possible. For some of the years, value is missing due to either new initiatives or the business unit was not able to quantify value for those years. It also contains data analytics investments (i.e. Capex and Opex). Capital expenses are shown as depreciation/amortisation costs. In some years the value shown is lower than the previous year due to foreign exchange conversion rate variations from local currency to dollar equivalent.

Table 23: Data analytics value – Proforma Profit and Loss statement

(Benefits $ Millions)	Data Analytics Value – Proforma Profit and Loss Statement							
	Year – 3	Year – 2	Year – 1	Current Year	Year + 1	Year + 2	Year + 3	Total
Benefits Value by Division and Functions								
Commercial Division								
Marketing	$38.06	$37.61	$39.63	$41.79	$52.97	$41.34	$50.20	$301.60
Pricing								
Sales and Distribution								
Commercial Division - Net Value	$38.06	$37.61	$39.63	$41.79	$52.97	$41.34	$50.20	$301.60
Customer Care								
Service and Customer Excellence			$0.04	$0.03	$0.02	$0.12	$0.28	$0.49
Customer Data Management	$2.48	$2.18	$2.08	$2.03	$1.88	$1.73	$1.68	$14.06
Customer Care - Net Value	$2.48	$2.18	$2.12	$2.06	$1.90	$1.85	$1.96	$14.55
Corporate Affairs								
Regulatory Compliance	$1.26	$2.33	$2.20	$1.39	$1.58	$1.44	$1.51	$11.72
Corporate Affairs - Net Value	$1.26	$2.33	$2.20	$1.39	$1.58	$1.44	$1.51	$11.72
Finance Division								
Business Planning and Analysis								
Financial Closing and Reporting	$1.16	$1.02	$0.98	$1.38	$1.28	$0.96	$0.93	$7.72
Risk and Revenue Assurance		$0.70	$0.00	$0.49	$0.16	$0.28	$0.41	$2.05
Taxation	$2.18	$2.33	$2.59	$3.24	$3.03	$3.15	$3.68	$20.20
Finance - Net Value	$3.34	$4.05	$3.58	$5.12	$4.47	$4.39	$5.03	$29.98
Infrastructure Division								
Energy and Power Infrastructure						$8.30	$8.08	$16.38
Field Operations								
Performance Analysis								
Infrastructure Planning						$2.50	$5.00	$7.50
Infrastructure Division - Net Value						$10.80	$13.08	$23.88
Financial Service Division								
Financial Services Sales & Marketing			$0.01	$0.08	$0.37	$0.69	$1.21	$2.36
Finance Services- Net Value			$0.01	$0.08	$0.37	$0.69	$1.21	$2.36
Total Benefits	$45.14	$46.17	$47.54	$50.44	$61.30	$60.51	$72.99	$384.09
Operating Expenses								
System Support		-$0.80	-$0.81	-$2.15	-$2.15	-$2.45	-$2.45	-$10.81
Application (Software) Maintenance	-$0.15	-$0.15	-$0.18	-$0.28	-$0.27	-$0.18	-$0.20	-$1.39
Other Expenses								
Depreciation Expenses	-$1.98	-$3.00	-$3.94	-$2.72	-$2.82	-$4.55	-$4.97	-$23.98
Total Operating Expenses	-$2.13	-$3.94	-$4.93	-$5.15	-$5.24	-$7.18	-$7.62	-$36.18
Tax Impact (Notional Tax)	-$15.06	-$14.78	-$14.91	-$15.85	-$19.62	-$18.67	-$22.88	-$121.77
Net Operating Profit After Tax	$27.96	$27.45	$27.70	$29.44	$36.44	$34.67	$42.49	$226.14

Table 24 and Table 25 are supplementary to Table 23 which illustrates ROI summary and how cash flows occur. Table 25 shows cash inflows representing the data analytics business benefits, and cash outflows as capital expenditures, and operating expenditures as they occur. The ROI summary and financial metric assumptions for the ROI calculations are also included in Table 24. It contains corporate tax as notional tax as well as weighted cost of capital together with depreciation rate used. The Net Present Value (NPV), IRR and payback period illustrate attractiveness of investing in data analytics.

Table 24: Data analytics Proforma ROI summary

Return on Investment Summary		**Financial Metrics**	
Net Present Value (NPV) [$ Millions]	$116.58	Weighted Average Cost of Capital	15.00%
Internal Rate of Return (IRR)	350.4%	Corporate Tax Rate	35.00%
Approximate Payback Period (Months)	3	Depreciation (Years)	5.55

Table 25: Data analytics Proforma Cash flows

Benefits $ Millions	Data Analytics Value – Proforma Cash Flow Statement								
	Year – 4	Year – 3	Year – 2	Year – 1	Current Year	Year + 1	Year + 2	Year + 3	Total
Benefits Value by Divisions and Functions									
Commercial Division									
Marketing		$38.06	$37.61	$39.63	$41.79	$52.97	$41.34	$50.20	$301.60
Pricing									
Sales and Distribution									
Commercial Division - Net Value		$38.06	$37.61	$39.63	$41.79	$52.97	$41.34	$50.20	$301.60
Customer Care									
Service and Customer Excellence				$0.04	$0.03	$0.02	$0.12	$0.28	$0.49
Customer Data Management		$2.48	$2.18	$2.08	$2.03	$1.88	$1.73	$1.68	$14.06
Customer Care - Net Value		$2.48	$2.18	$2.12	$2.06	$1.90	$1.85	$1.96	$14.55
Corporate Affairs									
Regulatory and Interconnect		$1.26	$2.33	$2.20	$1.39	$1.58	$1.44	$1.51	$11.72
Corporate Affairs - Net Value		$1.26	$2.33	$2.20	$1.39	$1.58	$1.44	$1.51	$11.72
Finance Division									
Business Planning and Analysis									
Financial Closing and Reporting		$1.16	$1.02	$0.98	$1.38	$1.28	$0.96	$0.93	$7.72
Risk and Revenue Assurance		$0.00	$0.70	$0.00	$0.49	$0.16	$0.28	$0.41	$2.05
Taxation		$2.18	$2.33	$2.59	$3.24	$3.03	$3.15	$3.68	$20.20
Finance - Net Value		$3.34	$4.05	$3.58	$5.12	$4.47	$4.39	$5.03	$29.98
Infrastructure Division									
Energy and Power Infrastructure							$8.30	$8.08	$16.38
Field Operations									
Performance Analysis									
Infrastructure Planning							$2.50	$5.00	$7.50
Infrastructure Division – Net Value							$10.80	$13.08	$23.88
Financial Service Division									
Financial Services Sales & Marketing				$0.01	$0.08	$0.37	$0.69	$1.21	$2.36
Finance Services- Net Value				$0.01	$0.08	$0.37	$0.69	$1.21	$2.36
Cash Inflow of Benefits		$45.14	$46.17	$47.54	$50.44	$61.30	$60.51	$72.99	$384.09
Cash Outflows									
Capital Investment									
Hardware Costs	-$7.04	-$4.61	-$5.06	-$3.95	$0.00	-$7.96	-$1.50	-$2.00	-$32.13
Application (Software) Costs		-$0.71	-$0.14	-$0.11	-$0.32	-$0.81	-$0.45	-$0.75	-$3.29
Development Costs		-$0.30	-$0.04	-$0.18	-$0.24	-$0.80	-$0.40	-$0.45	-$2.41
Total Annual Capital Investments	-$7.04	-$5.63	-$5.24	-$4.24	-$0.57	-$9.56	-$2.35	-$3.20	-$37.83
Operating Expenses									
System Support			-$0.80	-$0.81	-$2.15	-$2.15	-$2.45	-$2.45	-$10.81
Application Maintenance		-$0.15	-$0.15	-$0.18	-$0.28	-$0.27	-$0.18	-$0.20	-$1.39
Other Expenses									
Total Annual Operating Expenses		-$0.15	-$0.94	-$0.99	-$2.43	-$2.42	-$2.63	-$2.65	-$12.20
Taxes (Notional Tax)		-$15.06	-$14.78	-$14.91	-$15.85	-$19.62	-$18.67	-$22.88	-$121.77
Net Cash Flow	-$7.04	$24.32	$25.20	$27.39	$31.59	$29.70	$36.87	$44.26	$212.29

Many organisations use a variety of methods of funding data analytics investments. Table 26 provides a good example of centralised funding with cost charge back to the business divisions.

Cost charge back is calculated proportional to the extent of data analytics resources usage. This approach enables divisional level profitability of data analytics investments to identify which divisions are making the best use of data analytics to generate more value.

Table 26: Data analytics value – Divisional allocation of benefits and costs

(Benefits $ Millions)	Data Analytics Value – Divisional Allocation of Benefits and Costs								
	Year – 3	Year – 2	Year – 1	Current Year	Sub-total	Year + 1	Year + 2	Year + 3	Total
Commercial Division	**$38.06**	**$37.61**	**$39.63**	**$41.79**	**$157.09**	**$52.97**	**$41.34**	**$50.20**	**$301.60**
- Data Analytics Costs	-$1.32	-$2.45	-$3.00	-$3.14	-$9.92	-$3.20	-$4.38	-$4.65	-$22.15
- Data Analytics Net Benefits	**$36.73**	**$35.16**	**$36.63**	**$38.65**	**$147.16**	**$49.77**	**$36.96**	**$45.55**	**$279.45**
Customer Care	**$2.48**	**$2.18**	**$2.12**	**$2.06**	**$8.84**	**$1.90**	**$1.85**	**$1.96**	**$14.55**
- Data Analytics Costs	-$0.09	-$0.17	-$0.15	-$0.15	-$0.56	-$0.16	-$0.22	-$0.23	-$1.16
- Data Analytics Net Benefits	**$2.39**	**$2.01**	**$1.97**	**$1.90**	**$8.28**	**$1.74**	**$1.64**	**$1.73**	**$13.39**
Corporate Affairs	**$1.26**	**$2.33**	**$2.20**	**$1.39**	**$7.19**	**$1.58**	**$1.44**	**$1.51**	**$11.72**
- Data Analytics Costs	-$0.07	-$0.13	-$0.10	-$0.10	-$0.40	-$0.10	-$0.14	-$0.15	-$0.80
- Data Analytics Net Benefits	**$1.19**	**$2.21**	**$2.10**	**$1.29**	**$6.79**	**$1.48**	**$1.30**	**$1.36**	**$10.93**
Finance Division	**$3.34**	**$4.05**	**$3.58**	**$5.12**	**$16.09**	**$4.47**	**$4.39**	**$5.03**	**$29.98**
- Data Analytics Costs	-$0.51	-$0.95	-$1.13	-$1.18	-$3.79	-$1.21	-$1.65	-$1.75	-$8.40
- Data Analytics Net Benefits	**$2.83**	**$3.10**	**$2.44**	**$3.93**	**$12.30**	**$3.27**	**$2.74**	**$3.28**	**$21.58**
Infrastructure Division	**$0.00**	**$0.00**	**$0.00**	**$0.00**	**$0.00**	**$0.00**	**$10.80**	**$13.08**	**$23.88**
- Data Analytics Costs	-$0.13	-$0.24	-$0.25	-$0.26	-$0.88	-$0.26	-$0.36	-$0.38	-$1.88
- Data Analytics Net Benefits	**-$0.13**	**-$0.24**	**-$0.25**	**-$0.26**	**-$0.88**	**-$0.26**	**$10.44**	**$12.70**	**$22.00**
Financial Services Division	**$0.00**	**$0.00**	**$0.01**	**$0.08**	**$0.09**	**$0.37**	**$0.69**	**$1.21**	**$2.36**
- Data Analytics Costs	$0.00	$0.00	-$0.30	-$0.31	-$0.60	-$0.31	-$0.43	-$0.46	-$1.81
- Data Analytics Net Benefits	**$0.00**	**$0.00**	**-$0.29**	**-$0.23**	**-$0.52**	**$0.06**	**$0.26**	**$0.75**	**$0.55**
Total Data Analytics Gross Benefits	**$45.14**	**$46.17**	**$47.54**	**$50.44**	**$189.29**	**$61.30**	**$60.51**	**$72.99**	**$384.09**
Total Data Analytics Costs	-$2.13	-$3.94	-$4.93	-$5.15	-$16.14	-$5.24	-$7.18	-$7.62	-$36.18
Total Data Analytics Net Benefits	**$43.01**	**$42.23**	**$42.61**	**$45.29**	**$173.14**	**$56.06**	**$53.34**	**$65.37**	**$347.90**

Findings and recommendations

Besides the data analytics value statements, the report will need to be accompanied by key findings from the meetings along with recommendations for further improvements. The findings from the meetings will be measured against data analytics quality attributes as well as the organisation's capability ranking index which we will cover later in Chapter 21. In a nutshell, the report will contain status of best practices adoption for each business division that can be one of the following along with a description of why:

- Fully adopted
- Mature
- Some components of best practices in-place
- Need addressing

The report will also include areas that require improvements to drive more value.

CHAPTER 18

Data analytics value estimation

Overview: This chapter will discuss a framework for measuring and attributing data analytics value when several constraints arise in engaging with the stakeholders and managers. It leverages data analytics value benchmarks as a guide to achieve the desired outcome.

Data Analytics Value Estimation (DAVE)

Data analytics value estimation is a top-down approach to estimate the quantifiable business value of data analytics investments. It focuses on passive/semi-active engagement with stakeholders. It is used during the budgeting and planning process to develop the value driven data analytics for new opportunities. It can also be used during data analytics value assessment/audit process to quantify the benefits of data analytics investments using industry benchmarks, when the stakeholders and managers are unable to devote any time for meetings to confirm achieved value or discuss data analytics value contributions and/or they do not measure data analytics value and/or lack knowledge of an effective way to attribute data analytics value.

Figure 45 provides a high-level process for data analytics value estimation. It involves a review of the organisation's strategy and financials from published and/or internal reports to identify organisational strategy, goals and objectives. Industry research will also help in identifying the organisation's standing in the market and/or industry. Additionally, the organisation can identify issues/bottlenecks in current processes from the previous data analytics value assessment/audit activities, if available. These issues can be grouped into categories of operational, analytics, and technology problems that can help in developing estimates of opportunity costs and solution costs for including in the future road map. This is followed by developing estimates of value by business divisions and functions using benchmarks which we will discuss in the next few sections.

Review Organisation's Financial Report	Develop Estimation of Value	Determine Gaps and Identify Opportunities	Determine Potential Solutions and Costs
• Review organisation's strategy and financials from half-yearly / annual reports • Identify organisational strategy, goals / objectives	• Develop estimation of value by business division / function using benchmarks • Review with stakeholders and refine estimates	• Identify issues / bottlenecks in current processes and group these into categories of operational, analytics, technology problems • Develop estimates of opportunity costs	• Determine potential solutions and costs • Map future opportunities to organisational strategies which can be included in road map
• Published shareholder reports • Industry research from public sources / Analysts reports	• Benchmarks and estimation templates • Leverage value principles from Data Analytics Value Assessment method • Baseline references from previous estimates	• Leverage previous Data Analytics Value Assessments to identify opportunities and/or review published industry reports to compare the organisation's standing	• Leverage industry research on technology costs as well as previous Data Analytics Value Assessments

Figure 45: Data analytics value estimation process

Data analytics (AI/ML) value benchmarks

With the emergence of AI and Agentic AI, there is a spate of benchmarks being produced by organisations such as OpenAI's GDPVal benchmark[41] and Massachusetts Institute of Technology's (MIT) Iceberg Index.[42] For instance, GDPVal benchmark is designed to track how well AI models perform on economically valuable, real-world tasks performed by human experts.

MIT's Iceberg captures technical exposure, where AI can perform occupational tasks, not displacement outcomes or adoption timelines. Iceberg Index is a skills-centred KPI for the AI economy. It measures the wage value of skills that AI systems can perform within each occupation, revealing where human and AI capabilities overlap. It shows how technical capability extends far below the surface through cognitive automation spanning administrative,

41 https://openai.com/index/gdpval/

42 https://iceberg.mit.edu/

financial, and professional services.

While these benchmarks have their merits in evaluating the capabilities of AI technologies and their impact on human expert tasks, they are macro-economic benchmarks, focused on GDP (Gross Domestic Product), wages, unemployment, broader workforce skill development, and so on.

The scope of benchmark for this book is concerned with organisational level comparisons of the value contributions of data analytics (AI/ML) towards achieving business outcome relevant for their respective industries and business functions. Also, unlike the other AI benchmarks mentioned earlier, which exclusively focus on conversational AI, this book considers adoption of data analytics by many organisations over the last few decades in the form of BI, data mining, statistical predictions, machine learning, natural language processing (NLP) and the recent AI/ML for strategic and operational decision making.

There are different levels of benchmarks that are possible starting with industry level followed by organisational level, then by business division/functions. Corresponding to each level, the process for detailed estimation will vary. Besides different levels of value benchmarks used for data analytics value attributions, there are also benchmarks for data analytics costs which we will discuss.

Figure 46 illustrates data analytics value contributions as a percentage range (i.e. 1.2% to 12%) of the organisation's revenue as the basis for financial services, retail and telecommunication industries. This benchmark range is expected to expand in the near future as organisations continue to leverage advancements in Machine Leaning (ML) and Agentic AI solutions.

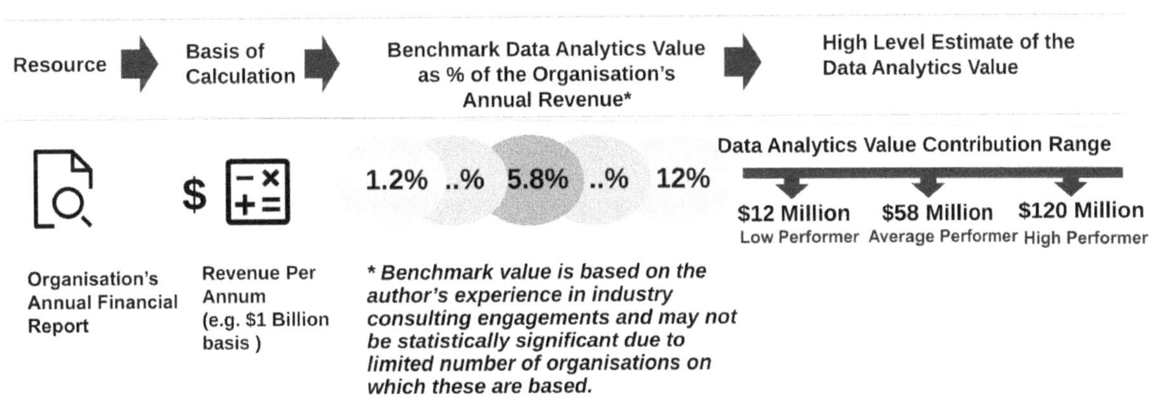

Figure 46: High level estimates of data analytics value based on all-industry benchmark

The benchmark percentage is a reflection of both enterprise-wide adoption of data analytics as well as the organisation's data analytics governance and management capability. Correspondingly, the data analytics value generated for enterprise-wide use of data analytics tends to be relatively higher. While enterprise-wide use in some industries or organisations may be high they do not automatically equate to higher value. A compounding factor in achieving higher value is driven by the individual organisational factors. Organisations with higher capability in data analytics governance and management are high performers. Assuming the annual revenue of the organisation is $1 billion, data analytics value for low performers can be estimated as $12 million, average performers as $58 million and high performers as $120 million.

While benchmarks for other industries have not been specifically mentioned, the above benchmarks can be used for organisations in other industries at the organisation level and for business division/function levels. Also, the benchmarks can also be used for 'not for profit' organisations, but their divisional levels will be different for these organisations. Since 'not for profit' organisations largely rely on source of 'income' as funds or grants received from government treasury and other institutions, in the place of revenue as the basis of data analytics value calculations, the total funding received, including any revenue earned by the organisation can be used as the basis of calculation. For example, in the health industry many government bodies receive funds or grants as their source of income but, they also earn revenues for many services they provide.

It should be noted that the all-industry benchmark data analytics value may not be statistically significant due to the limited number of organisations on which the benchmark values are based. However, the benchmarks are based on very large enterprises of around twenty-five organisations, with presence across the Australasia Pacific region. As mentioned earlier, there are varying levels of benchmarks that can differ based on each organisation's capabilities in managing and governing data analytics. Readers are advised to use their own judgement and discretion when using the reference benchmarks covered in this book.

Industry-specific data analytics (AI/ML) value benchmarks

Tables 27 and 28 contain data analytics value contributions by business divisions and functions for financial services, and telecommunication industries respectively. Two metrics for data analytics value contributions are shown (i.e. contribution as a % of total financial metrics and as a % of revenue respectively). These tables also contain KPIs relevant for the business functions within the industry. Financial metrics on which the KPIs are calculated are included on the basis of $1 billion annual revenue, and industry specific expenses proportional to the $1 billion revenue. For financial services industry, expenses and impairments are in the range of 45% to 60% of revenue, whereas for the telecom industry it is generally higher than 65% due to capital intensive infrastructure, and network investments and maintenance. Customer facing front-office business functions have a high proportion of revenue as their KPIs with some component of cost management. Other back-office business functions carry a high proportion of cost management with some revenue management in their job functions.

The tables provide aggregated data analytics value in millions of dollars based on the percentage of their contributions towards the respective business functions' effectiveness in achieving their business objectives and outcome. The individual business function's percentage contributions of data analytics vary subject to the extent to which these business functions rely on and use them in executing their job functions. The tables contain minimum, average and maximum percentage contributions of data analytics for each of the business functions by business division/department.

Table 27: Benchmark data analytics (AI/ML) values by business functions – Financial services

Business Divisions / Departments, Business Functions and Key Performance Indicators			Financial metric basis assumption for KPIs* ($m)	Percentage (%) of the business function's effectiveness attributable to Data Analytics (AI/ML)			Data Analytics Value ($m)
Business Division / Department	**Functional Areas**	**Key Performance Indicators (KPI)**		**Minimum**	**Average**	**Maximum**	
Consumer Banking	Sales and Service Marketing / Customer Management Product Management Pricing Management Network and Operations Credit Account, Savings Accounts, Transactions	Grow and retain revenue; customer life time value; new product revenue; optimise price; cost to acquire, serve and retain customers; customer experience and NPS.	$500	0.80%	4.96%	8.75%	$24.8
Business Banking	Similar functions as Consumer Banking but with managed accounts instead of mass market		$300		6.00%		$18.0
Institutional and Global Markets Banking Divisions	Corporate banking and foreign exchange services etc		$200		0.68%		$1.4
Chief Financial Office	General Ledger Reporting – Budgeting, Forecasting etc Margin, Working capital and Cash flow Management Regulatory Reporting Liquidity Management	Cost and profitability; Net Interest Margin; Revenue assurance; Asset and Liability management; Liquidity Coverage Ratio; Net Stable Funding Ratio; Impairment.	$285	4.00%	5.20%	6.75%	$14.8
Chief Risk Office	Credit Risk, Market Risk, BASEL 3 Regulatory compliance and reporting Operational Risk Portfolio Risk Financial Crime Prevention compliance and reporting	Expected Credit Loss; Probability of Default; Loss Given Default; Capital Adequacy Ratio; Asset and Liability Management;	$228	3.50%	9.34%	14.00%	$21.3
Information Technology and Data Analytics Management	Enterprise Analytics and Reporting Enterprise Services	Revenue Reporting; Total Cost of Ownership of IT; Data quality, integration and governance;	$57	5.00%	6.84%	15.00%	$3.9
***Note: Basis of annual revenue assumed as $1 billions; annual expenses and impairments of $570m**		**Total - - - >**	**$1,570**				**$84.17**
Data Analytics Value as a % Revenue + Expenses + Impairments (Average %) - >							**5.36%**
Data Analytics Value as a % of Revenue - - - - - >							**8.42%**

These tables also contain enterprise level data analytics value as a percentage of total financial metrics basis used including revenue, impairments and expenses in addition to percentage of revenue as its basis. For the financial services industry example in Table 27, it has 5.36% as data analytics value as a percentage of total financial metrics, and 8.42% as a percentage of revenue (i.e. $1 billion basis). The telecom industry example in Table 28 shows 6.20% as data analytics value as a percentage of total financial metrics, and 10.61% as a percentage of revenue (i.e. $1 billion basis).

Table 28: Benchmark data analytics (AI/ML) values by business functions – Telecom

Business Divisions / Departments, Business Functions and Key Performance Indicators			Financial metric basis assumption for KPIs* ($m)	Percentage (%) of the business function's effectiveness attributable to Data Analytics (AI/ML)			Data Analytics Value ($m)
Business Division / Department	Functional Areas	Key Performance Indicators (KPI)		Minimum	Average	Maximum	
Marketing and Product Management	Customer Value Management (CVM) Segmentation Product Portfolio and Pricing Optimisation Customer Insights	Grow and retain revenue; customer life time value; targeting accuracy; cross-sell and up-sell; Churn reduction; new product revenue; optimise price; Average Revenue per User (ARPU)	$385.5	6.00%	11.23%	20.00%	$43.3
Customer Management	Customer Experience Multi-channel Customer Care and Service	Cost to acquire, serve and retain customers; customer experience and NPS; first-time call resolution; call handling duration;	$306.5	4.00%	4.42%	7.50%	$13.5
Sales and Distribution	Channel Sales Prepaid, Postpaid, Recharges activation (Device, SIMS and PINS)	Sales revenue growth and retention; distribution costs and commissions; cost of sales.	$185.5	2.00%	3.08%	4.00%	$5.7
Networks (Mobile and Broadband) and Infrastructures	Network Planning and Operations Customer Level Network Analytics	Network capex and opex; network utilisation ratio; network coverage; network reliability and assurance; customers' network experience.	$284.0	1.00%	3.06%	3.75%	$8.7
Chief Financial Office and Corporate Service	Revenue Assurance (Billing & Collections) Risk and Audit (Fraud management) Lawful Enforcement Agency (LEA) reporting Revenue, Commissions, Interconnect Settlement, Budgeting and Forecasting Group Reporting Cost, Margin and Profitability	Cost and profitability; Revenue assurance; Capital and operating expenses ROI; Assets and liability; Collections; regulatory compliance.	$463.0	2.00%	5.78%	10.00%	$26.8
Information Technology and Data Analytics Management	Enterprise Analytics and Reporting Enterprise Services	Revenue Reporting; Total Cost of Ownership of IT; Data quality, integration and governance;	$85.5	2.00%	9.40%	12.00%	$8.0
*Note: Basis of annual revenue assumed as $1 billions; annual expenses of $710m		Total - - - >	$1,710				$106.06
Data Analytics Value as a % Revenue + Expenses + Impairments (Average %) - >							6.20%
Data Analytics Value as a % of Revenue - - - - - - - >							10.61%

The differences in data analytics value between the two organisations are dependent on a number of factors, such as the extent of data analytics usage in the respective business functions and jobs, industry specific business functions/data analytics requirements, and capability levels in data analytics management and governance. By reviewing the minimum, average and maximum percentages between the two tables for similar business functions we can notice differences. For instance, marketing and customer management functions for telecom in Table 28 shows higher percentages than for financial services in Table 27 but the reverse is the case for IT and data analytics management. The explanation for these differences emerges when meeting with the members of the business functional areas by asking questions. Additionally, the financial services industry has specific risk management functions (e.g. BASEL 3 compliance and financial crime prevention), and branch/ATM network management, and operation whereas, the telecom industry has Law Enforcement Agency (LEA) reporting, and mobile broadband network infrastructure management, and operation functions.

Dependencies in data analytics (AI/ML) value chain

On the aspect of data analytics capability levels, further considerations need to be given about the capabilities across the data analytics value chain. Figure 47 is a simplified view of the data analytics value chain. It illustrates the dependencies across data, analytics (AI/ML), insights, judgement/decision-making, actions taken and business performance. The value contributions of data analytics depend on the capabilities of managing and governing each of these components in the value chain.

Data in itself does not have any value. It derives its value by putting it to use in business activities and serves the utility function. Similarly, analytics in itself does not have any value. Analytics derives its value by being able to predict unknown information, and provide insights by using known data, and information as its source. Additionally, the value of data, and analytics-driven prediction further depends on the judgement made by the user as to their quality, accuracy, and reliability in making decisions by considering any risks involved in their use. The measure of business performance, and outcome from actions taken by the user and/or Agentic AI application generally determine the extent of usefulness of data analytics, and associated value. This underscores the need for effective management, and governance of data and analytics in order for their reliability, and trustworthiness for use in the business.

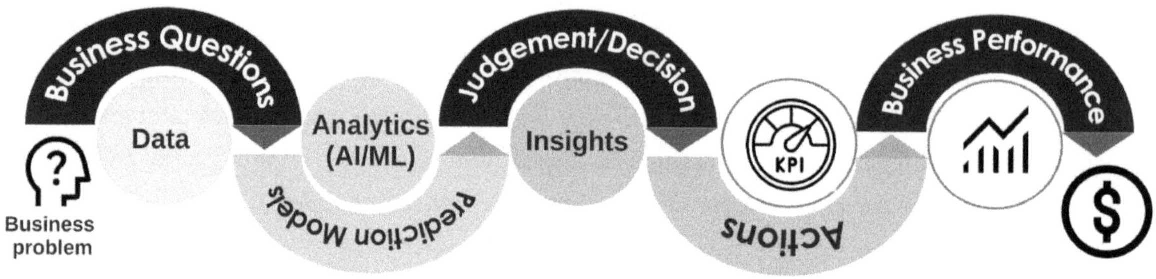

Figure 47: Dependencies in data analytics (AI/ML) value chain

How to use the data analytics (AI/ML) value benchmarks

As discussed earlier in this chapter under 'Data Analytics Value Estimation (DAVE)', the benchmarks can be used as part of the high-level process for data analytics value estimation. The benchmarks are useful for establishing baseline value of data analytics when the stakeholders and managers are unable to quantify its value.

Figure 45 provides a high-level process for data analytics value estimation. It contains a step to review the organisation's strategy, and financial statements from published and/or internal reports to identify organisational strategy, goals and objectives. The financial statements can be used for developing estimates of data analytics value by business divisions, and functions using the benchmarks as described below.

Two approaches can be used for estimating data analytics value. One is based on the industry-specific average using the industry examples from Table 27 (i.e. 5.36% of total financial metrics) or Table 28 (i.e. 6.20% total financial metrics), and the other is by using all-industry averages (i.e. 5.8% of revenue) by using Figure 46. The difference is that the former industry-specific value contribution uses total financial metrics consisting of revenue, expenses and impairments as the basis, whereas the latter all-industry average, uses revenue as the basis for estimation of value. However, each of the approaches can additionally use the resulting value estimates of data analytics to calculate its contribution on the basis of revenue and total financial metrics of the organisation respectively.

Table 29: Data analytics (AI/ML) value estimate based on Telecom industry average

Business Divisions / Departments, Business Functions and Key Performance Indicators			Basis of benchmark values		Financial metric* for this organisation	Estimate based on Telecom industry average benchmark - Business function's effectiveness attributable to Data Analytics	
Business Division / Department	**Functional Areas**	**Key Performance Indicators (KPI)**	**Financial metric basis assumption for KPIs* ($m)**	**Average %**	**Business function's financial metric for KPIs ($m)**	**Percentage (%) contribution**	**Value contributions ($m)**
Marketing and Product Management	Customer Value Management (CVM) Segmentation Product Portfolio and Pricing Optimisation Customer Insights	Grow and retain revenue; customer life time value; targeting accuracy; cross-sell and up-sell; Churn reduction; new product revenue; optimise price; Average Revenue per User (ARPU).	$385.5	11.23%	$3,528.11	11.23%	$396.28
Customer Management	Customer Experience Multi-channel Customer Care and Service	Cost to acquire, serve and retain customers; customer experience and NPS; first-time call resolution; call handling duration.	$306.5	4.42%	$2,805.10	4.42%	$123.92
Sales and Distribution	Channel Sales Prepaid, Postpaid, Recharges activation (Device, SIMS)	Sales revenue growth and retention; distribution costs and commissions; cost of sales.	$185.5	3.08%	$1,697.70	3.08%	$52.26
Networks (Mobile and Broadband) and Infrastructures	Network Planning and Operations Customer Level Network Analytics	Network capex and opex; network utilisation ratio; network coverage; network reliability and assurance; customers' network experience.	$284.0	3.06%	$2,599.18	3.06%	$79.62
Chief Financial Office and Corporate Service	Revenue Assurance (Billing & Collections) Risk and Audit (Fraud management) Lawful Enforcement Agency (LEA) reporting Revenue, Commissions, Interconnect Settlement, Budgeting and Forecasting Group Reporting Cost, Margin and Profitability	Cost and profitability; Revenue assurance; Capital and operating expenses ROI; Assets and liability; Collections; regulatory compliance.	$463.0	5.78%	$4,237.40	5.78%	$245.00
Information Technology and Data Analytics Management	Enterprise Analytics and Reporting Enterprise Services	Revenue Reporting; Total Cost of Ownership of IT; Data quality, integration and governance;	$85.5	9.40%	$782.50	9.40%	$73.58
***Note: Annual revenue for this organisation $9.5 billions; annual expenses of $6.15 billions**		Total - - - >	**$1,710**	**6.20%**	**$15,650**	**6.20%**	**$971**
Data Analytics Value as a % Revenue + Expenses (Average %) - - - - >						**6.20%**	
Data Analytics Value as a % of Revenue - - - - - - - - - >						**10.22%**	

Table 29 illustrates this approach by using the telecom industry average percentage contribution of 6.20% from Table 28 as the basis for the organisation under consideration that has annual revenue of $9.5 billion, and expenses of $6.15 billion as reported in the organisation's financial statements. This table provides financial metric basis for KPIs as well as detailed value contributions by business divisions/departments and business functions proportional to benchmark values. The data analytics value contributions of $971m from Table 29 can also be used to derive its contributions by revenue which is 10.22%.

Table 30 describes the alternative approach by using the all-industry average percentage contribution of 5.8% of revenue from Figure 46 as the basis for the organisation under consideration that has annual revenue of $9.5 billion, and expenses of $6.15 billion as reported in the organisation's financial statements. Table 30 shows the financial metric basis for KPIs as well as detailed value contributions by business divisions/departments and business functions proportional to benchmark values. The data analytics value contributions of $551m in Table 30 can also be used to derive its contributions by revenue and expenses which is 3.52%.

It should be noted that in Table 29 and Table 30, the financial metrics for KPIs ($m) for each business division/function are aggregated from several individual financial metrics associated with the KPIs (i.e. Grow Revenue; Retained Revenue; Avoid Costs; Reduce Costs; Cash flow; Profitability, etc).

While not included as separate tables above, a similar approach as the telecom examples can be used for a financial services organisation that has annual revenue of $8.85 billion, expenses of $4.75 billion, and impairments of $320 million as reported in the organisation's financial statements. The data analytics value contributions are $746m based on financial services average data analytics value contribution of 5.36% based on total financial metrics. Its value contributions based on revenue are 8.43%.

Table 30: Data analytics (AI/ML) value estimate based on all-industry average

Business Divisions / Departments, Business Functions and Key Performance Indicators			Basis of benchmark values		Financial metric* for this organisation	Alternative estimate based on Al Industry average benchmarks (5.8% of Revenue) - Business function's effectivenes attributable to Data Analytics	
Business Division / Department	**Functional Areas**	**Key Performance Indicators (KPI)**	**Financial metric basis assumption for KPIs* ($m)**	**Average %**	**Business function's financial metric for KPIs ($m)**	**Percentage (%) contribution**	**Value contributions ($m)**
Marketing and Product Management	Customer Value Management (CVM) Segmentation Product Portfolio and Pricing Optimisation Customer Insights	Grow and retain revenue; customer life time value; targeting accuracy; cross-sell and up-sell; Churn reduction; new product revenue; optimise price; Average Revenue per User (ARPU).	$385.5	11.23%	$3,528.11	6.38%	$224.95
Customer Management	Customer Experience Multi-channel Customer Care and Service	Cost to acquire, serve and retain customers; customer experience and NPS; first-time call resolution; call handling duration.	$306.5	4.42%	$2,805.10	2.51%	$70.34
Sales and Distribution	Channel Sales Prepaid, Postpaid, Recharges activation (Device, SIMS)	Sales revenue growth and retention; distribution costs and commissions; cost of sales.	$185.5	3.08%	$1,697.70	1.75%	$29.66
Networks (Mobile and Broadband) and Infrastructures	Network Planning and Operations Customer Level Network Analytics	Network capex and opex; network utilisation ratio; network coverage; network reliability and assurance; customers' network experience.	$284.0	3.06%	$2,599.18	1.74%	$45.20
Chief Financial Office and Corporate Service	Revenue Assurance (Billing & Collections) Risk and Audit (Fraud management) Lawful Enforcement Agency (LEA) reporting Revenue, Commissions, Interconnect Settlement, Budgeting and Forecasting Group Reporting Cost, Margin and Profitability	Cost and profitability; Revenue assurance; Capital and operating expenses ROI; Assets and liability; Collections; regulatory compliance.	$463.0	5.78%	$4,237.40	3.28%	$139.07
Information Technology and Data Analytics Management	Enterprise Analytics and Reporting Enterprise Services	Revenue Reporting; Total Cost of Ownership of IT; Data quality, integration and governance;	$85.5	9.40%	$782.50	5.34%	$41.77
***Note: Annual revenue for this organisation $9.5 billions; annual expenses of $6.15 billions**		**Total - - - >**	**$1,710**	**6.20%**	**$15,650**	**3.52%**	**$551**
Data Analytics Value as a % Revenue + Expenses (Average %) - - - - >						**3.52%**	
Data Analytics Value as a % of Revenue - - - - - - - - - >						**5.80%**	

Incremental values of data analytics (AI/ML)

The previous section illustrated the approaches for estimating data analytics value using benchmarks. These approaches are useful for establishing baseline value of data analytics it currently generates for the enterprise across business divisions and functions. It can also be useful for some organisations who are starting from scratch and are interested in understanding potential value of data analytics for the organisation by considering industry norms.

The data analytics usage is expected to continue. Therefore, its current value will continue to accrue into future years with any adjustments for changes in business requirements, and corresponding data analytics usage. The need for adjustments occurs due to changes in KPIs and/or changes in business applications and/or changes in regulatory requirements and/or changes in market dynamics. These adjustments will form part of a road map and corresponding incremental value of data analytics which will be aggregated to recurring future values.

Changes to business requirements are identified when reviewing the data analytics value estimations with business representatives, and/or as part of the data analytics value assessment. These requirements are described as use cases in a standard template for consistency. An example of the use case can be found in Table 31.

Table 31: Data analytics use case example – Complaints Process Optimisation

Complaints Process Optimisation	
When a customer complains the issue needs to be resolved quickly and efficiently to prevent significant impact on the relationship and costs to the organisation. Automatically identify from unstructured texts and conversations originating from multiple channels (e.g. Online, Call Centres, e-mail, Mobile App) to systematically categorise and group what customers are complaining about for prioritising and early closure.	
Problem Statement / Objectives	**Challenges**
1. Customer complaints and feedback about products and services from multiple channels are not coordinated well; 2. Some customers interact in their native language requiring translations. Challenges arise in consistently documenting the severity and seriousness of the complaint or feedback due to unstructured text format that leads to potential loss or misunderstanding; 3. There are inconsistencies in the individual agent's complaint handling performance that leads to inefficiencies and costs.	Multi lingual translations; Unstructured data; Quality of text data.
Expected Outcome When Objectives Met	**Data Required for Business Outcome**
1. Early identification and resolution of complaints before systemic problem; 2. Prioritisation of expert resource on most important cases; 3. Improved customer satisfaction from ability to highlight emerging service issues and take action; 4. Root cause analyses to address process bottlenecks; 5. Understanding of individual complaint handler performance.	1. Customer profile; 2. Complaint / feedback descriptions and logs from all channels (voice and texts); 3. Agent data on which agent(s) are dealing with the complaint; 4. Current and historical outcome of complaint / feedback along with date and time of resolution; 5. Large Language Models (LLM).
Business Benefits of Data Analytics	**Analytical Methods for Business Outcome**
KPIs: Improve Meantime for Complaint Resolution by 10%; Reduce Costs by 3% to 6% and Improve Retained Revenue by 5% 1. Auto categorisation of inbound complaints routing to relevant agents to improve efficiency; 2. Understand true cost of complaint handling and rework by complaint type; 3. Anticipate reasons for complaint and proactively offer relevant recommendations; 4. Dynamic optimisation of resources and costs.	1. Retrieval Augmented Generation (RAG) and natural language processing (NLP); 2. Latent Dirichlet Allocation (LDA) for topic modelling of complaints / feedback; 3. Sentiment analyses to determine customer reactions; 4. Cosine Similarity on complaint descriptions to identify and anticipate for knowledge management.

The use case contains a title, brief description, problem statement or use case objectives, expected outcome, business benefits of the use case along with KPI measures, some challenges encountered, data required for the use case and analytical methods applicable for addressing the problem and/or business objectives. A use case does not have a predetermined business value, but this can be estimated by using the KPI improvements sought, and associated baseline financial metrics of revenue, expenses, and other costs which will provide the incremental benefits when the use case is implemented as a solution.

Here is a description of the data analytics value behind the use case in Table 31 that can be calculated from detailed analyses from the various meeting notes, and reports in the organisation. Let's say the analyses found the following facts:

- Increase in net income compared with previous year is $125.3 million for the telecom organisation shown in Table 29.
- Out of the total of $125.3 million, the Customer Management division has been responsible for the net income increase of $22.52 million. Data analytics value contributions of this increase can be calculated from Table 29 (i.e. 4.42%) as $995,384. This amount represents the value contributions of data analytics as a cumulative addition to the previous year due its continued use in augmenting job tasks performed by the customer management division. Assuming we are doing data analytics value assessment during the current year, this amount is already included as shown in Table 29. Now, we are interested in developing a road map of future data analytics value from use cases to be included in subsequent years. The use case in Table 31 is one of the many use cases to be considered for future years (i.e. Year + 1, Year + 2, Year + 3, etc).
- From the analyses, we note that the organisation is currently achieving customer retention rate of 7.7% by using data analytics, and the skills of the customer management division to prevent churn of 7.7% of customers. Customer retention is the responsibility of all customer-facing front-line divisions. So, Customer Management division's proportion of the financial metrics for net retained revenue stands at $1.73 million. The use case is targeting 5% improvement on the customer retention rate which translates to $86,500. Data analytics value contributions towards this financial metric can be estimated from Table 29 (i.e. 4.42%) as $3,823.
- Our analyses also find that of the many customer-initiated calls and messages to the customer management team during the year, 211,600 were related to customer complaints with a total cost of $5.29 million to the organisation. The use case is targeting to reduce this cost by 3% to 6%. On average, this would amount to a cost reduction of $317,400. Data analytics value contributions are estimated as $14,029.
- The mean time for complaint resolution is currently about 18 minutes which the use case is targeting to reduce by 10%. The target improvement of 1.8 minutes is already reflected in the above improvements in cost reduction and retained revenue estimations. Hence, not separately calculated.

- The value contributions of data analytics for the use case in Table 31 can be estimated as $17,852 (i.e. $3,823 + $14,029). This use case is one example of the many business initiatives that may be taking place in parallel in the organisation, which will form the basis of road map discussions based on the magnitude of data analytics value contributions of each use case, business priority, data availability for the use case, resource/skills availability, and so on.

The incremental benefits from the number of use cases undertaken each year along with reduction of any data analytics use that are decommissioned in a year will provide aggregated net incremental value for subsequent years, together with recurring annual value of its contribution to the business outcome.

It should be noted that in addition to new use cases, there are many use cases that the organisation will continue to use for many months and years to come. Use cases such as customer acquisition, churn reduction, cross-sell, up-sell, pricing optimisation, new product introduction, and predictive maintenance are among the list of common use cases in regular use. However, these use cases will mostly likely be based on revised KPI targets aligned to the organisation's yearly revision of business strategy and planning. In addition, the use cases will need fine tuning to improve prediction accuracy by integrating additional data sources, and updates to analytical algorithms. These changes will require re-estimation of the data analytics value of the recurring use cases by taking into account the revisions from the previous versions.

Another point to note is that while benchmark values are a useful and effective way to estimate data analytics value contributions, we can expect drifts in the benchmarks due to the revisions of recurring use cases, and new use cases being introduced. As a result, the benchmarks will need to be reset once every two or three years by performing data analytics value assessment.

Data analytics (AI/ML) benchmark costs and return on investments

So far, we have discussed approaches for estimating data analytics value by using value benchmarks. In order to measure ROI, we also need data analytics costs. While costs are relatively easy to obtain and measure compared with data analytics value, we often stumble into difficulties caused by the reluctance of business representatives to share data analytics costs. Many organisations outsource components of IT, and data analytics investments and/or use varying mixes of on-shore and off-shore resources for a variety of tasks that make it difficult estimate costs. It was found that the concerned business representatives are also not transparent about costs. Additionally, many organisations do not report IT costs, and data analytics costs in their annual reports, while a handful of organisations do report. From the many customer engagements, we have been able to compile benchmarks of costs for financial services and telecom industries.

Information technology (IT) costs or spend for the financial services industry ranges from 7% to 10% of revenue, whereas for the telecom industry it is 5% to 7%. Data analytics spend for financial services ranges between 18% and 25% of IT spend; for the telecom industry it is between 17% and 21% of IT spend.

Table 32 presents the data analytics costs for financial services and telecom industries along with ROI for the example financial services and telecom organisations previously covered. After allowing for notional tax on benefits, the median ROI for the financial services organisation is 261%; for the telecom organisation it is 557%. The ROI range, not shown in the table, for financial services organisation is 166% to 399%; for the telecom organisation the ROI range is 417% to 771%. These ROI ranges correspond to the data analytics costs, which are in the range of 18% to 25% for financial services, and 17% to 21% for telecommunications in the reverse order of ROI (i.e. when cost is high, ROI is lower).

Table 32: Industry data analytics (AI/ML) costs and ROI

Industry Data Analytics Costs and Return on Investments (ROI)								
Industry	Revenue ($m)	Data Analytics Value ($m)	IT Costs ($m)			Data Analytics Costs ($m)		
			Low	Median	High	Low	Median	High
Financial Services organisation	$8,850	$746						
- IT Spend as % of Revenue			7.00%	8.50%	10.00%			
- IT Spend			$620	$752	$885			
- Data Analytics Spend as % of IT Spend						18.00%	21.00%	25.00%
- Data Analytics Spend						$112	$158	$221
		Median						
- Gross benefits		$588						
- Notional Tax on benefits (30%)		$177						
- Net benefits		$412						
- Simplified Return on Investments (ROI)		261%						
Telecom organisation	$9,500	$971						
- IT Spend as % of Revenue			5.00%	6.00%	7.00%			
- IT Spend			$475	$570	$665			
- Data Analytics Spend as % of IT Spend						17.00%	19.00%	21.00%
- Data Analytics Spend						$81	$108	$140
		Median						
- Gross benefits		$862						
- Notional Tax on benefits (30%)		$259						
- Net benefits		$604						
- Simplified Return on Investments (ROI)		557%						

CHAPTER 19

Data analytics value optimisation

Overview: The main focus of this chapter is in describing methods and techniques for sweating the data analytics investments by optimising their use and doing more use cases with under-utilised data resources.

Data Analytics Value Optimisation (DAVO)

Data analytics value optimisation is a bottom-up approach to identify business value opportunities that can be exploited using readily available data in the organisation's data lakehouse ecosystem. It also helps to identify under-utilised data, and leverage data reuse opportunities to drive additional business value from industry use cases.

Why data reuse matters?

Organisations invest in capturing and storing large amounts of data from their internal systems and external sources to support operational and management planning, and decision-making. On average it costs $21,915 for acquiring and integrating each new data source that goes through the process of refinement, enrichment and quality assurance using the organisation's extract, transform and load process (ETL), and information engineering principles of data normalisation. A small-sized project with 30 data sources costs $657,450 for integrating the data into the data lakehouse ecosystem. A large organisation will have hundreds, if not thousands, of such data sources integrated into their multi-petabyte size data lakehouse.

Unless this investment is put to effective use, it becomes a sunk cost. Among the contributing factors of sunk costs are departmental and application-oriented data stores that inhibit value. Besides this, the division between business and technology departments compounds this problem by becoming an obstacle that further inhibits value. The internal data analytics providers in IT departments understand the data but have limited knowledge of what business use cases can be supported with the data. On the other side, the business users have ideas of what business problems they want to address but have poor visibility of what data is currently available in the organisation, and what analytics are needed. Opportunities always exist to generate additional value from the data readily available in the organisation. The challenge is to quickly and efficiently identify available data that can be exploited for business use cases.

How data reuse drives more business value

Data reuse unlocks this potential to create business value by identifying business-outcome-driven use cases that exploit the data already available in the data analytics ecosystem. Data reuse can be defined as the use of the data for multiple use cases beyond the use case for which it was originally intended. Data reuse allows an organisation to do more with less by leveraging data already in the data analytics ecosystem and limiting new investments to the integration of only incremental new data. Reuse of data reduces wasteful investments, while increasing trustworthiness, as it encapsulates and assumes many of the best practices in data governance and management. Since the data source has previously gone through the process of refinement, enrichment, and quality assurance using the organisation's extract, transform and load process, it can be considered reliable for reuse.

Challenges in data reuse

Technology reuse is the easiest and has the least impact on business outcome. Data, and the related analytics application reuse have greater impact on business outcome but are harder to achieve. Simply assuming that the data elements in the dataset can be used for new analytics and solutions as they are does not work! This is because the synonyms, homonyms and semantic meaning of the data elements, and the context of analytics application or solutions will need to be understood well and standardised for their reuse. This is especially true for Agentic AI applications that leverage the emerging Model Context Protocol (MCP) Model Context Protocol (MCP), Knowledge Graph and ontology to provide AI agents with complete semantic richness of enterprise data and meaningful context.

Consider the following examples of data elements in the financial services industry-related terminologies:

- 'bank' implies a financial institution or edge of a river
- 'current' in relation to an account represents customer's current account for day-to-day transactions, and electricity in the utility industry context
- 'net' in the context of financial statement could mean net profit after expenses, or a mesh in telecommunications context
- 'fine' indicates financial penalty for breach of regulatory compliance, or appreciation of good service provided by a front-office agent
- 'book' as in closing the books represents finalising all transactions for a defined period on a cut-off date, or to make an appointment with a bank's representative for a mortgage application.

Similarly, consider Figure 48 that contains the different semantic meanings attached to the data element 'Call Duration' that represents the time duration of a customer making an international call. The answers to the question, '*How long was this international call?*' has different meanings to the concerned departments.

- For the network department, call duration is calculated for the entire period from when the customer dials the number + ringing duration + call answer time by the recipient + duration of conversation until hangup + time to release the circuit. This duration represents for how long the telephone circuit or line was occupied, which is a useful measure for forecasting circuit capacity needed in network planning.
- For the billing department, call duration is calculated as the duration of useful conversation from when the recipient answers the call + duration of conversation until hangup + any rounding off adjustments as per the tariff.
- For the marketing department, call duration is calculated as the duration of useful conversation from when the recipient answers the call + duration of conversation until hangup.

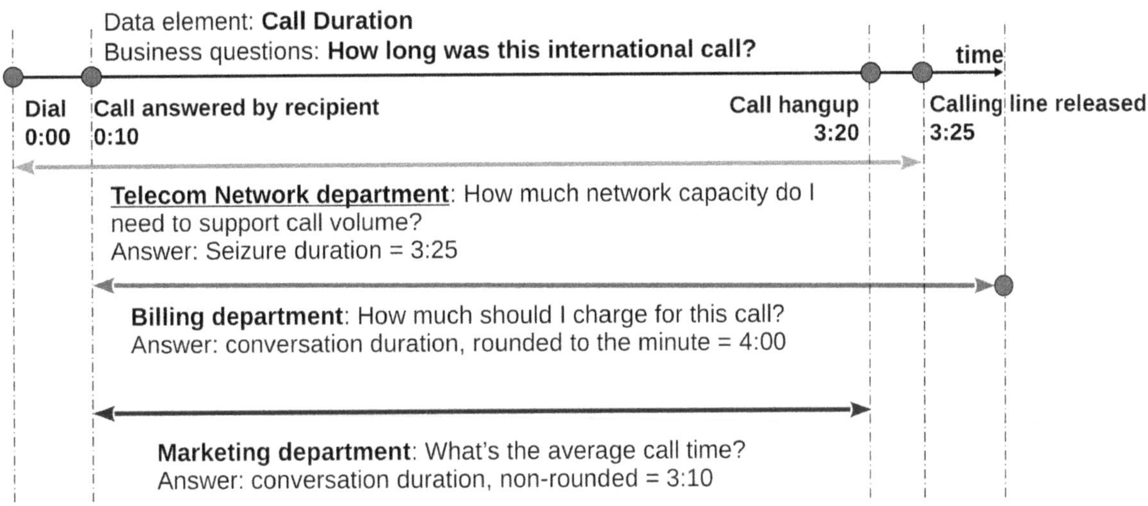

Figure 48: Telecom industry example – semantic meaning of data for international call duration

This is important because, making simple assumption that 'Call Duration' is reusable across any use case without its semantic meaning, and context is misleading. A network specific use case will require total duration of the call during which the circuit was fully occupied including network signalling time + conversation time. A billing use case will require for what duration of the call the customer was charged with an applicable tariff, and a marketing use case will require duration of actual useful conversation period.

Likewise, consider another example of data that relates to membership of a customer in social networks, such as Facebook, Instagram, or LinkedIn. A marketing use case that wants to leverage this social network membership of the customer for cross-sell or up-sell campaigns will be interested in strength of connections and influence in the network, known in advanced analytics as Eigenvector Centrality. Whereas, in a customer experience management use case, the interest will be on the messages and comments communicated by the customer in the social network with peers to understand the sentiment of the customer by taking advantage of natural language processing (NLP) analytics.

Data management and standardisation for data reuse

When business users define their requirements for a use case, it is generally in the form of free-flowing descriptive texts. Besides this, each business user may describe the texts in the domain specific terminology, and in their own preferred style. These constraints lead to inconsistencies in definitions of data elements required for processing, for predictions using advanced analytics, for calculations of performance metrics, and reporting. Naturally, these prevent re-usability, and associated benefits that could otherwise be derived from reuse. In order to take advantage of data reuse, the data will need to be managed effectively by means of standardisation of metadata, and associated context, and rules.

Let us review the use cases from different industries/business functions in Table 33 to Table 35 that contain data requirements for the use cases.

Table 33: Supply Chain use cases with data requirements

Use Case No.	Use Case Title	Data Required for Business Outcome
1	Supply Chain Financial Performance Analysis	1. Operational purchases, shipments, costs and sales 2. Labour hours and rates 3. Transportation and manufacturing costs 4. Inventory 5. Master data (exchange rates, contracts, etc)
2	Supply Chain Finance – Order to Cash	1. Sales Orders 2. Contracts 3. Payment, tax, freight and invoice information 4. Pricing (sales) conditions 5. Customer/Product Master Data 6. Financial data

Both use cases in Table 33 are related to finance in supply chain business function. However, the focus of the use cases and their respective data requirements are different. The objective of use case 1 is about overall effectiveness of the supply chain business function. The data requirements are focused on total costs and revenue from all activities (i.e. purchase, shipment, sales, transportation, inventory holding, supplier contracts, labour hours, etc) involved in the supply chain business function with an expected business outcome of gross margin and profitability. Whereas the objective of use case 2 is to identify working capital requirements by focusing on how quickly sales orders can be converted into cash (i.e. realised revenue).

Both use cases mention sales, contract and master data in their requirements. However, their semantic meanings and data element requirements are very different. For instance, 'Sales' in use case 1 refers to revenue whereas, in use case 2 'Sales' refers to sales order, selling price, conditions of sales, data of sale for tracking time between sales order, invoice, freight shipment and payment. Similarly, 'Contract' in use case 1 refers to supplier contract whereas, in use case 2 it refers to sales contract or condition of sales. Master data between the use cases are different. In short, for reusing the data elements from these two use cases a clear definition of the data items along with any synonyms, and homonyms, and standardisation of metadata will be required.

Table 34: Customer Management use cases with data requirements

Use Case No.	Use Case Title	Data Required for Business Outcome
3	Customer Interaction Management - Multi-step Campaigns	1. Event triggers and customer listening agents to trigger contact opportunities / measure success 2. Detailed closed loop outcomes from channels to trigger stepping of campaign activity 3. Campaign management functionality to orchestrate customer journey steps and automate campaign execution 4. Contact policy framework to control solicitation
4	Customer Life Cycle Management - Customer Journey Leading to Churn	1. Data Warehouse for customer profile, customer value, package subscription, churn status 2. Billing records Call centre data - call centre logs 3. IVR data 4. e-Service logs – customer session history 5. Predicted churners list from existing predictive models (for comparison purpose)
5	Customer Interaction Management - Omni-Channel Customer Engagement	1. Channel usage summary data 2. Internet Banking transaction logs 3. Mobile Banking App logs including geo-location data 4. Branch usage data 5. Telephony usage logs 6. Transactional Data – cross channel 7. Activity Based Costing transaction data 8. Product Holdings 9. Customer Demographics 10. Customer Geo-demographics 11. Customer Profitability

All the use cases in Table 34 relate to monitoring and interpreting the life cycle activities of customers in order to take appropriate actions for managing effective interactions with customers. However, the business outcome, and data requirements of each use case, are varied and different. Use cases 3 and 5 focus on management strategy for multi-step campaigns, and multi-channel (also known as 'omni-channel') customer engagements from aggregation of collective customer activities, whereas use case 4 is for profiling customers, and predicting customer events (also known as 'customer journey steps') leading to potential customer churn such that churn prevention, and customer retention actions can be taken. Interestingly, there are common data items between the three use cases that are described differently. Outcome from channels in use case 3, e-Service logs, call centre logs, IVR (Interactive Voice Recorder) data in use case 4, and channel usage in use case 5 all refer to customer interaction events across different channels. Use case 5 expands the data items relating to channel usage by individually listing digital, and bricks-and-mortar channels, viz. online, mobile, telephone, and branch visits. Again, they reinforce the need for better data definition management and standardisation of terminologies.

Table 35: Network Assets use cases with data requirements

Use Case No.	Use Case Title	Data Required for Business Outcome
6	Asset Life Cycle Management - Improve Network Operation to Minimise Asset Ageing	1. Asset budget and financial 2. Asset re-enforcement plans 3. Network design / specification 4. Network topology 5. Core asset register 6. Metered consumption 7. SCADA alarms Asset telemetry feeds 8. Generation asset profiles 9. Consumer consumption profiles 10. Asset loading reports 11. Weather feeds 12. Asset operation reports 13. City data Building data 14. Demographic trends
7	Service Assurance - Customer Network Experience Analytics	1. Probe systems – 3G, 4G, 5G RAN and Core 2. DPI data PM/FM data 3. Reference data Handset info (TAC codes) 4. Customer data 5. Churn risk 6. Complaint history

The use cases 6 and 7 in Table 35 relate to network infrastructure assets from two different industries. Use case 6 is concerning electricity generation, transmission and distribution network, whereas use case 7 is for telecom mobile networks. The use cases use different terminologies to describe the networks. In use case 6, the data 'Generation asset profile' denotes electricity generation, whereas in use case 7 the references to '3G, 4G, and 5G' stand for different generations of mobile network evolution; 'RAN and Core' stand different kinds of networks in telecom, viz., Radio Access Network and Central Office network respectively. In use case 6, 'SCADA alarms Asset telemetry feeds' refers to logs of electricity network events. In use case 7 the equivalent is known as 'DPI data' (i.e. deep packet inspection data); 'PM' for network performance management data; 'FM' for fault management data.

In short, both use cases are interested in analysing network events data logs for various purposes, including network utilisation and performance to ensure the networks are providing high service quality and customer experience, reduce health and safety risks, reduce carbon emissions, and protect the environment. These use cases also illustrate the need for consistent data management, standardisation and governance.

Best practices in data management and governance dictate that conceptual modelling, logical modelling, and semantic modelling of data are required to be performed using appropriate standards prevalent in the organisation before physical design and development of the databases, and data storage activities. These standards may involve the principles of information engineering discipline, such as entity-relationship modelling and/or object-relational modelling. The benefits of logical and semantic modelling include effective linkage in the end-to-end translation of business requirements into technical design and implementation. This also helps in providing freedom of technology choice (e.g. data warehouse, data lakehouse, data mesh/data fabric, object data store) to the technical teams when implementing the business requirements.

Table 36: Supply Chain use cases with standard data terms

Use Case No.	Use Case Title	Standard Data Item Term	Data Subject Domain Term
1 and 2	Supply Chain Financial Performance Analysis Supply Chain Finance - Order to Cash	customer master financial agreement product master product price supplier contract terms supplier inventory supplier invoice supplier master supplier performance supplier production supplier purchase order supplier transit time supply chain constraint supply chain cost supply chain inventory supply chain purchase order supply chain transaction sales invoice sales order	Customer - Profile Customer - Account Product - Catalogue Product - Price Supplier - Contract Supplier - Inventory Supplier - Transaction Supplier - Profile Supplier - Performance Supplier - Inventory Supplier - Transaction Supplier - Transaction Supplier - Capacity Supplier - Profitability Supplier - Inventory Supplier - Order Supplier - Transaction Sales - Transaction Sales - Transaction

Tables 36 to 38 contain two levels of data standardisation of the data requirements in the use cases in Table 33, Table 34 and Table 35 respectively. 'Standard data item term' refers to first level of standardisation, and 'Data subject domain term' refers to a higher level of grouping of 'Standard data item term'. This allows for data reuse at different levels as required. The advantage of higher-level data reuse is that it abstracts reuse to less than 200 data terms which may be useful and helps with '*seeing the woods for the trees*', whereas the lower levels run into hundreds of standard data item terms. It also helps with drilling down to lower level when needed.

Table 37: Customer Management use cases with standard data terms

Use Case No.	Use Case Title	Standard Data Item Term	Data Subject Domain Term
3,4 and 5	Customer Interaction Management - Multi-step Campaigns Customer Life Cycle Management - Customer Journey Leading to Churn Customer Interaction Management - Omni-Channel Customer Engagement	campaign control group campaign target customer channel preference customer churn event customer life stage customer lifetime value customer marketing communication event customer marketing communication event outcome customer mobile app interaction event customer mobile banking transaction customer mobile interaction event customer offline interaction event customer online interaction event customer online interaction event outcome customer online real-time interaction event customer privacy preference customer product return customer product review customer risk score customer sentiment customer service interaction event customer service performance customer status customer transaction	Marketing - Targeting Marketing - Targeting Customer - Profile Customer - Event Customer - Profile Customer - Profile Customer - Event Customer - Event Customer - Digital Channel Customer - Transaction Customer - Digital Channel Customer - Offline Channel Customer - Digital Channel Customer - Digital Channel Customer - Digital Channel Customer - Profile Customer - Product Customer - Product Customer - Profile Customer - Sentiment Customer - Offline Channel Customer Service Customer - Profile Customer - Transaction

Table 38: Network Assets use cases with standard data terms

Use Case No.	Use Case Title	Standard Data Item Term	Data Subject Domain Term
6 and 7	Asset Life Cycle Management - Improve Network Operation to Minimise Asset Ageing Service Assurance - Customer Network Experience Analytics	asset capacity	Asset - Profile
		asset configuration	Asset - Profile
		asset cost	Finance - Balance sheet
		asset coverage span	Asset - Profile
		asset demography	Asset - Profile
		asset design specification	Asset - Profile
		asset failure event	Asset - Event
		asset failure reason	Asset - Event
		asset life cycle cost	Asset - Profitability
		asset loading report	Asset - Profile
		asset maintenance cost	Asset - Profitability
		asset maintenance log	Asset - Maintenance
		asset operation status	Asset - Profile
		asset performance measurement	Asset - Performance
		asset performance specification	Asset - Performance
		asset repair event	Asset - Event
		asset risk profile	Asset - Profile
		asset safety report	Asset - Regulatory
		asset utilisation	Asset - Transaction
		asset valuation	Asset - Profitability
		customer churn risk score	Customer - Event
		customer profile	Customer - Profile
		customer mobile network signalling event	Customer - Event
		customer network experience measure	Customer - Sentiment
		customer network interaction event	Customer - Event
		customer network probe event	Customer - Event

Quantifying the business value of data reuse

The benefits of data standardisation that we discussed in the previous section will help in driving greater business value from data reuse. The best way to illustrate this is by using examples from the reports which a software generates. The software, which the author developed, provides a visual interface to the user to asking him/her to enter the organisation's annual revenue, expenses, and cost of goods sold. It also allows the user to select from a list of a large library of use cases that have potential for alignment with the organisation's objectives. The organisation's financial data entered by the user is used to estimate business value of data analytics use cases. Let's say the user selects the use cases which we discussed in Table 33, Table 34 and Table 35 from the list of use cases displayed on the visual interface. Based on the information entered by the user, the software generates several reports which we will discuss below.

The software provides a summary report by business functions as shown in Table 39 that contains the count of use cases selected by the user. The report also contains the count of use cases by business function that reuse the data from the selected use cases. We can see from Table 39 that the data from 7 selected use cases creates expanded analytical opportunities for the organisation by reusing the data to solve 40 more business problems via 40 additional use cases. This represents enormous cost savings. Additionally, it is also expected to contribute to a potential analytical value of $21.84 million for the organisation (see Table 40). The business value of data analytics is estimated based on the organisation's financial information entered by the user along with additional factors that are built into the software, such as growth rate of the industry in the economy in which the organisation operates, and data analytics capability ranking of the organisation. However, the actual estimate of benefits will depend on the specific KPI metrics that the organisation wants to improve by using the suggested use cases.

Table 39: Summary count of use cases by business functions with data reuse

INDUSTRY RELEVANT CRITICAL BUSINESS USE CASES		
Business Function	**Selected Use cases for the Industry**	**Use cases that reuse data from selected use cases**
Finance		10
Infrastructure	2	8
Marketing / Customer Management	3	9
Operation		6
Product		4
Risk		
Security/Fraud		
Supply Chain	2	3
Total use cases	7	40

An interesting observation to be noted in Table 39 is that the use cases that reuse the data spread to other business functions beyond the business functions for which the seven use cases were selected by the user. For instance, the business functions of finance have ten use cases, operations management has six use cases, and product management four use cases. Besides, the supply chain, marketing and infrastructure business functions also contain additional use cases of one, six and six respectively that reuse the data. This illustrates the value of integrated data. Details on the quantity of data reuse for each use case are covered in Table 41 to Table 43 which we will discuss shortly.

Table 40: Data analytics value of use cases by business functions

ESTIMATED ANALYTICAL VALUE BY BUSINESS FUNCTION	
Business Function	**Estimated Analytical Value ($m)**
Finance	**$4.48m**
Infrastructure	**$5.88m**
Marketing / Customer Management	**$4.67m**
Operation	**$2.53m**
Product	**$1.85m**
Risk	
Security/Fraud	
Supply Chain	**$2.43m**
Total Estimated Analytical Value	**$21.84m**

The software that generated Table 39 containing count of use cases by business functions also generates the list of use cases in Table 41, Table 42 and Table 43 with their title, number of data subjects required for each use case, what percentage of the data subjects is already available in the data analytics ecosystem, what percentage of the data required for the use case is readily available through data reuse, and the percentage of new data required, if any.

Table 41: Supply Chain and Finance Management use cases with quantity of data reuse

SUPPLY CHAIN MANAGEMENT	DATA SUBJECTS Needed	Available
End to End Supply Chain Visibility	6	**50%+ 33% Reuse**
Cost to Serve Analysis	5	**20%+ 60% Reuse**
Supply Chain Financial Performance Analysis	8	**43%+ 57% New**

SUPPLY CHAIN MANAGEMENT	DATA SUBJECTS Needed	Available
Optimise On-Shelf Availability	4	**50%+ 50% Reuse**
Supply Chain Finance – Order to Cash	10	**50%+ 50% New**

FINANCE MANAGEMENT	DATA SUB-GROUPS Needed	Available
Budget and Forecasting Process Improvement	7	**71%+ 29% Reuse**
Billing and Receivable Analytics	6	**67%+ 17% Reuse**
Accounting Hub / Sub-ledger	8	**50%+ 50% Reuse**
Accelerated Financial Data Acquisition	5	**40%+ 40% Reuse**
Financial Reporting Efficiency and Effectiveness	6	**25%+ 75% Reuse**

FINANCE MANAGEMENT	DATA SUB-GROUPS Needed	Available
Sales Reporting and Analytics	9	**44%+ 56% Reuse**
Profitability Analytics	6	**67%+ 17% Reuse**
Data-driven Activity Based Costing	7	**43%+ 43% Reuse**
Executive Information Hub	7	**43%+ 57% Reuse**
Working Capital Management	8	**100% Reuse**

Arrow marks in the table denote the use case that the user selected. The software identifies what percentage of data required for the selected use cases are already available, and how much data are to be newly acquired. Table 41 shows that the 'Supply Chain Financial Performance Analysis' use case requires eight data subjects of which 43% of data is already available, and only 57% of new data is to be acquired. Similarly, the use case, 'Supply Chain Finance – Order to Cash' has 50% data subject already available, and only 50% of new data needs to be sourced.

The rest of the use cases in Table 41 contain quantities of what percentage of data is available, and how much of data is reused. Some use cases such as 'Working Capital Management' reuse 100% of the required data which implies cost savings in data acquisition, and integration for the project that delivers the use case.

Table 42: Marketing and Product Management use cases with quantity of data reuse

MARKETING MANAGEMENT	DATA SUBJECTS Needed	Available
▶ Customer Life Cycle Management - Customer Journey Leading to Churn	7	75%+ 25% Reuse
Event Triggered Offers	5	60%+ 20% Reuse
Real Time Personalisation	5	60%+ 20% Reuse
Micro Location Customer Targeting	6	50%+ 50% Reuse
▶ Customer Interaction Management - Multi-step Campaigns	8	50%+ 50% Reuse
Utilisation and Experience Based Segmentation	7	29%+ 57% Reuse
High Value Customer Retention	5	60%+ 20% Reuse
▶ Customer Interaction Management - Omni-Channel Customer Engagement	7	60%+ 20% Reuse
Location Based Marketing	6	50%+ 50% Reuse
Channel Optimisation - Customer Engagement	7	40%+ 40% Reuse
Customer Behaviour Analysis	5	40%+ 40% Reuse
High Value Customer Network Experience Analysis	8	40%+40% Reuse

PRODUCT MANAGEMENT	DATA SUBJECTS Needed	Available
Path to Change of Service	6	67%+ 17% Reuse
Product Retention Analytics	5	50%+ 50% Reuse
Behavioral Based Pricing with Telematics Data	7	33%+ 67% Reuse
Sales Team Analytics: Commissions	6	50%+ 50% Reuse

Similar to the Supply Chain and Finance use cases illustrated in Table 41, the Table 42 and Table 43 contain lists of use cases, number of data subject required, available and reused percentages for Marketing and Product Management, and Infrastructure and Operations Management business functions respectively.

In the tables with a list of use cases, it will be noticed that the available data subject percentages do not add to 100% for some use cases. This is because the software has the option of reporting only those use cases that have 80% or more data readily available for considerations and excludes those use cases that have less than 80% of readily available data.

Table 43: Infrastructure and Operations Management use cases with quantity of data reuse

INFRASTRUCTURE MANAGEMENT	DATA SUBJECTS Needed	Available
Service Assurance - Customer Network Experience Analytics	7	50%+ **50% Reuse**
Trunk Network Infrastructure Optimisation	6	33%+ **67% Reuse**
Assess Potential Performance vs. Manufacturer Specifications	5	20%+ **60% Reuse**
Quality Life Cycle Analysis	5	20%+ **60% Reuse**
Value-based Asset Decommissioning	5	20%+ **60% Reuse**

INFRASTRUCTURE MANAGEMENT	DATA SUBJECTS Needed	Available
Improve Maintenance Quality	6	33%+ **67% Reuse**
Service Path Load Patterns	4	25%+ **75% Reuse**
Optimise Design/Configuration of New Technology	5	20%+ **60% Reuse**
Early Warning Failure and Reliability	4	**100% Reuse**
Asset Life Cycle Management - Improve Network Operation to Minimise Asset Ageing	6	17%+ **83% New**

OPERATION MANAGEMENT	DATA SUBJECTS Needed	Available
Lease Optimisation	6	67%+ **33% Reuse**
Energy Cost Reduction	3	33%+ **67% Reuse**
Field Resource Utilisation	7	**100% Reuse**

OPERATION MANAGEMENT	DATA SUBJECTS Needed	Available
Fleet Funding Management	6	50%+ **50% Reuse**
Carbon footprint Reduction	5	10%+ **80% Reuse**
Proactive Reliability Centred Maintenance	5	20%+ **80% Reuse**

The rationale behind this is that for those use cases where there is a shortfall in available data it may considered justifiable to invest a small fraction of up to 20% of new data for these use cases.

The software that generates the data reuse tables also creates detailed pages for each of the use cases with a brief description, problem statement or use case objectives, expected outcome, business benefits of the use case along with KPI measures, data required for the use case, and analytical methods applicable for addressing the problem and/or business objectives as shown in Table 44. The software also produces other reports containing the list of KPIs applicable for the industry, as well as few other summaries. These reports have not been included here. The authors intend to make the software accessible to readers of the book as an online supplement to a future edition of this book.

In summary, data reuse is a very powerful technique for optimising data analytics investments by doing more with less, and by leveraging the investments that the organisation

Table 44: Customer Interaction Management – Omni-Channel Customer Engagement

Customer Interaction Management - Omni-Channel Customer Engagement	
Insurers and banks with multiple distribution channels need to know which channels, their customers, agents and brokers prefer to communicate and transact their business. Leverage insights to develop strategy to optimise channels to better meet customer needs.	
Problem Statement / Objectives	**Challenges**
Insurers and banks with multiple distribution channels are blindsided about the communication preferences of their customers, agents and brokers which prevent them from meeting the needs of their customers. There is a need for developing a strategy for effective customer engagements.	Creation of key channel behaviour and micro-segmentation
Expected Outcome When Objectives Met	**Data Required for Business Outcome**
1. Creation of key channel behaviour 2. Micro-segmentation 3. Transactional usage segmentation 4. Channel value segmentation 5. Channel usage segmentation 6. Customer value segmentation 7. The development of propensity models to predict channel 8. Recommend alternative channel	1. Channel usage summary data 2. Internet Banking transaction logs 3. Mobile Banking App logs incl. geo-location data 4. Branch usage data 5. Telephony usage logs 6. Transactional Data – cross channel 7. Activity Based Costing transaction data 8. Product Holdings 9. Customer Demographics 10. Customer Geo-demographics 11. Customer Profitability
Business Benefits of Data Analytics	**Analytical Methods for Business Outcome**
KPIs: Improve customer retention rate by 5%; Improve customer experience and NPS by 20%; Optimise channel spend by 5% 1. Insight led strategy to optimise channels to better meet customer needs 2. Enhanced user experience for key processes and customer segments 3. Increased customer satisfaction	1. Path Analyses with Sankey diagram visualisation 2. Attribution 3. Clustering 4. Collaborative Filtering 5. Optimisation

has already made. This has been previously illustrated in Chapter 10 under the heading 'Data leverage saves costs and speeds up time to value', which showed how project efforts, time and costs can be substantially reduced with data reuse and leverage. It is a sensible thing to do as part of best practices in data management and governance.

Analytics algorithm reuse

While data is reusable across many use cases other than the one for which the data was initially intended, the same cannot be said to be true for analytics algorithm reuse. This is because analytics algorithms behave differently on different datasets causing different results. Besides this, prediction models use a combination of analytics algorithms to improve the accuracy of prediction. In addition, prediction models decay over time due to changes in data on which the models were initially trained, which compounds the problem of analytics reuse for many use cases.

Let's consider advanced text analytics in NLP that often compare the similarity of two strings for various use cases. Most common similarity algorithms involve Damerau-Levenshtein distance, Jaro-Winkler distance, Needleman-Wunsch distance, Latent Dirichlet allocation (LDA) using Term Frequency (TF), Inverse Document Frequency (IDF), Cosine Similarity, Smith-Waterman, and Monge-Elkan and others to compute similarity scores. The author has used these algorithms for various business applications, such as detection of identity fraud, subscription fraud, topic modelling, sentiment analyses, and part of speech tagging for identifying names of people, and places in large text documents to build network of connections in syndicates. It was found that each algorithm behaves differently on the type of texts. Some algorithms, such as Jaro-Winkler, works better on proper names, whereas Damerau-Levenshtein works better on a person's contact details, such as phone number, and email addresses to check if these are manipulated, which is useful in fraud prevention applications.

In short, analytics algorithm reuse is not straightforward and requires considerable analyses and thought before reusing in various use cases.

Part IX

Data analytics (AI/ML) value governance and capability ranking

In Part IX we will discuss the role of value governance with its emphasis on periodic value audit, assessment, and communication by leveraging the data analytics value framework. Often the data analytics value achieved by an organisation correlates with its effectiveness in governance and data analytics capabilities management. In a separate chapter we will also discuss approaches, processes for data analytics capability assessment, and ranking which will help organisations in improving long-term value of data analytics investments.

CHAPTER 20

The need for value governance

Overview: This chapter describes the need for data analytics value governance and provides methods and processes for ensuring its effectiveness.

Data analytics value governance

One of the distinguishing characteristics of the role of data is that it has given rise to knowledge-based assets in improving organisational processes, business models, and has contributed to enhancing productivity, and economic growth. Hence, data analytics requires the same oversight as other tangible and intangible assets that are managed in the organisation.

Governance and stewardship

Data analytics is created and used across the enterprise. Therefore, governance encompasses cross-functional collaboration between business units and IT. One of the purposes of governance is to facilitate management-level cooperation among business and technology units to initiate data analytics strategies, and ensure alignment with business objectives, goals, and outcome while also contributing to integration and trust across the enterprise.

Stewardship enables representation of business operations that leverage data analytics use in business functions, activities, and software solutions. It is important that data analytics stewards represent industry and business knowledge, and domain expertise so that multiple business areas can share the data analytics assets effectively in decision-making to drive value. They should also ensure consistent definitions, and classifications of data, data quality, reuse, security, and trust. It is also important to ensure effectiveness of predictive model performance in model training, testing, and deployment besides monitoring for model decay.

Governing principles

Unlike traditional data analytics management practices, digital transformation brought with it rapid rate of data generation at enormous scale. While this provides great opportunities for organisations, there is also the need for safeguarding the data analytics assets due to the potential threat in collecting, processing, refining, storing, and using it. The key principles in data analytics value governance are not very different from data governance[43] and data science

43 TM Forum (2022), 'GB1023 Data Governance Guidebook v3.0.0' (01 June 2022).

governance, and it must be managed and governed as part the organisation's existing data management, and data science governance processes. These key principles consist of:

- **Top-down engagement**: Senior leadership must drive the data analytics strategy execution.
- **Growth mindset**: Digital transformation has demonstrated that data will continue to grow faster than our ability to manage it. This calls for the organisation to identify, prioritise, and implement data analytics governance, and stewardship mechanisms by selectively prioritising business use cases aligned with the organisation's strategic objectives.
- **Fit for purpose**: Ensuring that data analytics users' expectations are met, based on the intended use in delivering business outcome.
- **Positive outcomes**: Measuring data analytics' financial impact, productivity gains and compliance risks.
- **Best practices**: A set of rules, and best practices that govern how the organisation can produce, manage, and use data analytics assets for strategic and operational decision-making across the enterprise.
- **A value-driven process**: Measuring and managing quantifiable tangible and notional benefits of data analytics in supporting business process improvements, and business decisions. Data creates value. So, it should be protected and managed from a business, and from an accountability, and social responsibility perspective.
- **Audit**: Metadata governance provides an understanding of data analytics assets. It helps to determine the assets an organisation owns, their use across the enterprise, and provides the opportunity for tracking beneficiaries of value as well as enables mitigation of any risks.

  As part of value governance, periodic reviews, and audits provide opportunity for tracking, and communicating the realisation of data analytics investments to senior executives in the organisation.

Governance oversight

Data analytics value governance defines processes, procedures, and an execution plan about how people, process, and technology enable value. Steering Committees provide oversight to value governance activities, and ensure they are executed under the established guidelines, and best practices. Stakeholders will be from all areas of the business and IT. As the governance practice matures, the original strategy may need to change and continuously improve to accommodate new business drivers.

Stewards are responsible and accountable for value governance activities. Value governance activities will involve oversight of the points that we previously discussed throughout this book and which can serve as guidelines and steps to ensure that data analytics investments are providing long-term value. Key considerations for oversight are that:

- Data analytics investments are aligned with the organisation's strategic goals, and objectives referenced in Chapter 6, 'Maximise value by aligning with business strategy'.
- The value contributions of data analytics are effectively attributed by considering the resource mixes in achieving business outcome referenced in Chapter 7, 'Principles of data analytics value attribution'.
- Data analytics is supporting decision-making for business functions and activities by integrating with business applications across the enterprise, referenced in Chapter 8, 'Value chain in enterprise business functions'.
- Ensuring value governance oversight that considers data analytics value foundation as well as tracks value inhibitors and takes advantage of 'Data leverage with data reuse analyses' covered in Chapter 10.
- Reviewing the effectiveness of data analytics in augmenting job tasks as well as benefits of automation that deliver efficiencies: referenced in Chapter 12, under the heading 'Mapping of task characteristics and skill requirements with human and AI/ML'.
- Automation may lead to potential substitution of human tasks and will need the oversight of human resources managers to ensure employees' skills training and transfer of roles. A review of Chapter 13, under 'Automation and autonomous decision-making with AI' will help to understand the process of knowledge formation, observational learning, judgement, and decision-making to determine a strategy for investment in decisions relating to people, and emerging Agentic AI technologies by considering the risks and rewards.

  Besides this, Chapter 14, 'Techno-human dynamics in decision-making' provides insights on techno-human integration by evaluating the human psychological principles of attention, emotion, and reasoning.
- Conducting data analytics value assessment to track realisation of ROI by using data analytics value framework discussed previously by using the various methods, processes and tools covered.

- As mentioned previously due to the extended time and efforts needed for data analytics value assessment, the frequency for this audit can be once a year or once every two or three years. However, a more frequent short assessment or evaluation will be required once every three months to review, and record benefits of data analytics realised by using the benefits diary template shown in Table 45 that contains some examples related to the telecom industry. There is an associated Table 46 that contains drop-down lists of benefits process areas of the enterprise, benefits category, and benefits reliability which can be used for each project. The lists can be updated according to the organisation's industry, and domain requirements.
    - In Table 45, 'Project NPS' contains benefits realised by three different business process areas, three different benefits categories, and three different benefits stakeholders from across the enterprise.
- The steering committee can take advantage of Data Analytics Value Estimation (DAVE) (see Chapter 18) by using benchmark values during annual data analytics planning, and budgeting cycles. However, we can expect drifts in the benchmarks due to the revisions of input data and/or prediction models introduced in recurring use cases as well as new use cases. As a result, the value oversight team will need to revise and update the benchmarks as part of the periodic data analytics value assessment/audit.
- In order to take advantage of data reuse, the data will need to be managed effectively by means of standardisation of metadata, and associated context as described in Chapter 19.

Table 45: Data analytics benefits diary template

Project Name	Time period	Business Process	Benefit Category	Key Benefit Stakeholder	Brief summary of benefits attributable to Data Analytics (AI/ML)	Notional / Tangible	Planned $	Achieved $	Supporting Documents Location	Notes
Project NPS	From mm/yyyy to mm/yyyy	Customer Experience	Revenue Growth and Protection	Customer Experience Management	Proactive Bill Shock management that enables management of credit risks and customer experience while optimising network investments.	Notional				
		RevCost: Credit Risk Reduction	Cost Reduction	Billing Operations		Tangible				
		NW: Network Planning	Cost Avoidance	Network Operations		Tangible				
Project ClawBack		RevCost: Revenue Assurance	Revenue Protection (Retained Revenue)	Finance Operation and Control	Reconcile billed usage of each customer and usage from network to identify leakage.	Tangible				

Table 46: Drop-down lists for use in data analytics benefits diary template

Business Process Areas	Benefits Category	Benefits Reliability
Compliance: Contract	Cost Avoidance	Notional
Compliance: Financial	Cost Avoidance & Reduction	Tangible
Compliance: Police and Justice	Cash Flow Management	
Compliance: Regulatory	Cost Reduction	
Customer Acquisition	Profitability	
Customer Experience	Revenue Growth	
Customer Growth	Revenue Growth and Cost Avoidance	
Customer Retention	Revenue Growth and Cost Reduction	
Customer: Campaign Management	Revenue Growth and Protection	
DataMgmt: Enterprise Architecture	Revenue Protection (Retained Revenue)	
DataMgmt: Maturity	Revenue Protection and Cost Avoidance	
DataMgmt: Metadata Management	Revenue Protection and Cost Reduction	
NW Operations Management		
NW Route/Path Management		
NW Traffic Analysis		
NW: Network Planning		
NW:Network Performance		
RevCost: Access Cost Management		
RevCost: Content Assurance		
RevCost: Credit Risk Reduction		
RevCost: Enterprise Performance Management		
RevCost: Fin Op Analysis & Reporting Enhancement		
RevCost: Fin Reporting & Analysis Improvement		
RevCost: Fraud Prevention		
RevCost: Revenue Assurance		
RevCost: Settlements Assurance		
RevCost: Working Capital Optimisation		

Last but not least, effective value governance leads to higher levels of capability ranking and ensures that the organisation is a high performer. In the next chapter, we will discuss data analytics capability assessment and ranking in detail with industry examples.

CHAPTER 21

Capability assessment and ranking

Overview: In this chapter, we will extend data analytics value governance to encompass capability assessment of the organisation and associated ranking. The purpose of this is obvious when we observe high performing organisations who have also attained higher ranking in their data analytics management and governance capabilities.

Data analytics (AI/ML) capability assessment and ranking

Data analytics (AI/ML) capability assessment and ranking defines how an organisation can evolve in the management, governance, and use of data analytics to improve the organisational processes and decision-making to achieve desired business outcome, and long-term value from investments in data analytics. It includes key dimensions of oversight covered in 'Data analytics value governance' (Chapter 20) and consists of effectiveness in:

- aligning data analytics investments with organisation's strategic goals and objectives
- attributing value contributions of data analytics by considering the resource mix in job tasks
- the role of data analytics in supporting decision-making by business functions and activities across the enterprise by means of integration with enterprise applications
- ensuring that all aspects of data quality and data reuse are fit for purpose and serve as foundations of value to the organisation
- each step of the prediction model development process, and remediation of any biases that can occur during the analytics process
- determining strategies for investing in people and Agentic AI technologies by considering the risks and rewards and
- tracking realisation of value and return on investments on a periodic basis.

Five levels of progress in capability ranking for each of the above key dimensions can be considered:

Establish: Start by defining and developing policies for risk appetite and rewards for data analytics investments.

Make-do: Internal processes established with no cross-functional collaboration and integration.

Govern: Integrate cross-functional/enterprise-wide strategies and applications to drive synergies across data analytics management. Develop minimal standards, and governance framework, and audits for regulatory compliance and risks.

Adapt: Develop metrics to measure process performance and attribution of benefits. Adapt to changing market dynamics and regulatory requirements, and develop improvement strategies, but with minimal viable compliance.

Lead: Management innovation drives industry standards, leading best practices, and business models with a focus on continuous improvement in data analytics value contributions and risk management.

Table 47 contains example of the dimensions and ranking that can be used to assess and rank data analytics capabilities. The table shows current ranking and goal state for each dimension being evaluated. It also contains priority focus to achieve the goal state.

Table 47: Data analytics (AI/ML) capability assessment and ranking example template

	Dimension	Current Ranking	Goal State	Priority	Establish	Make-Do	Govern	Adapt	Lead
					1	2	3	4	5
A.	**Business and Technology Alignment**								
	1. Business Perspective								
	1.1. Business strategy, goals and objectives	1	3	H	—	—	▶		
	1.2. Business process improvements and business questions	1	3	H					
	1.3. Key performance indicators and metrics	2	4	H		—	—	▶	
	1.4. Cross-functional integration of business operations	1	3	M					
	1.5. Business decisions and actions	2	3	M					
	1.6. Enterprise and business functional performance tracking	1	4	M	—	—	—	▶	
	2. Technology Perspective								
	2.1. Enterprise data ecosystem architecture	1	3	M					
	2.2. Enterprise data ecosystem integration and automation	1	4	H	—	—	—	▶	
	2.3. Enterprise data ecosystem operations management	1	3	M					
	3. Human Resource and Cognition Perspective								
	3.1. Management team's sponsorship and direction	2	4	H		—	—	▶	
	3.2. Employees' skills, business and technical expertise	1	3	M					
B.	**Data Management and Governance**								
	1. Logical and semantic data modelling	1	3	L					
	2. Metadata management	0	2	M					
	3. Master data management	1	3	L	—	—	▶		
	4. Physical data modelling and design	1	3	M					
	5. Data quality attributes	1	4	H					
	6. Data accessibility	1	2	M					
	7. Security and Privacy	0	3	H					
C.	**Data Science Management and Governance**								
	1. Business understanding	0	3	M	—	—	▶		
	2. Data understanding	1	4	H					
	3. Data preparation	1	3	L					
	4. Prediction modelling	1	4	M					
	5. Evaluation	0	3	L					
	6. Deployment	1	3	M					
D.	**Data Analytics Value Tracking**								
	1. Data analytics value governance	0	2	L					
	2. Data analytics value assessment	1	4	M	—	—	—	▶	
	3. Data analytics value reporting and communication	1	2	L					

In preparing the data analytics capability assessment report, a chart as in the example in Figure 49 can be used to show the status of current ranking and goal state for the organisation. Beside the overall ranking chart, separate charts can be created for each of the major dimensions.

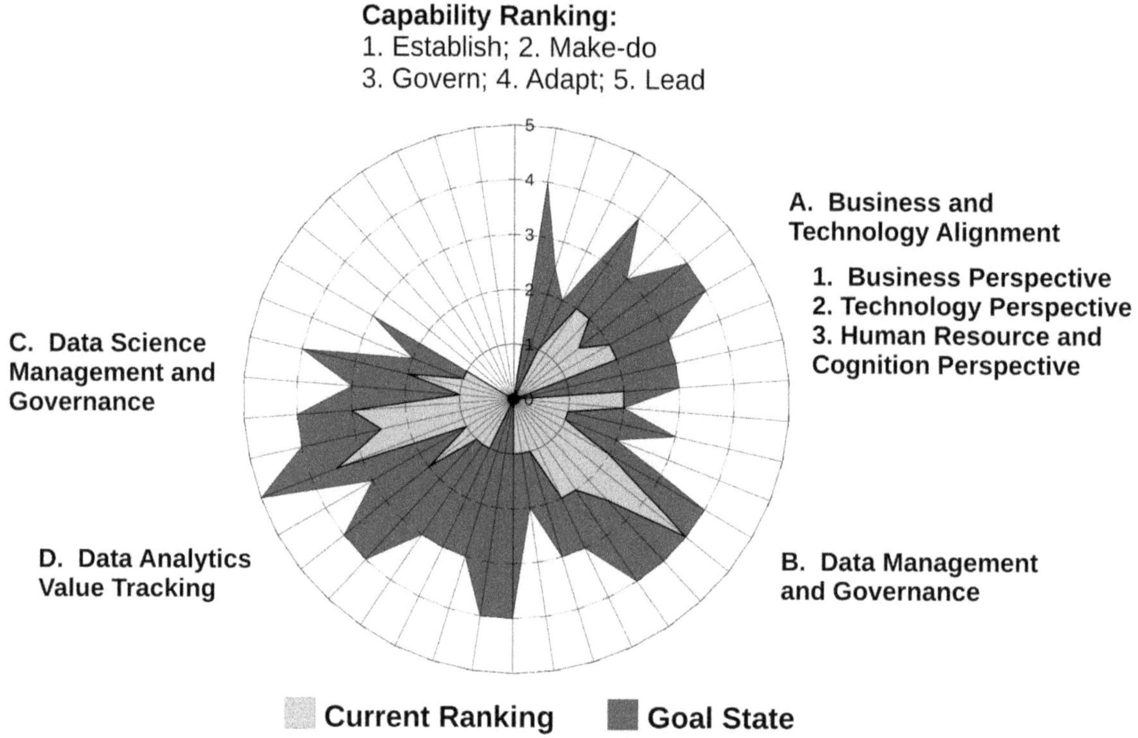

Figure 49: Data analytics (AI/ML) capability ranking chart example

While not explicitly covered in the data analytics capability assessment and ranking, the roles of project planning and management and deployment, production, operation, and maintenance by leveraging DataOps, ModelOps, AnalyticsOps are equally important and must be considered as part of the overall governance. If not managed and governed well they will impact data analytics value.

Part X

Illustrative case studies and industry use cases

This part of the book will illustrate, through case studies for financial services and telecom industries, some of the key topics discussed in this book that influence and contribute to data analytics value. They also show factors that inhibit data analytics value which need to be managed well by establishing capability assessment and governance.

Chapter 22

Illustrative case study of financial services industry

Overview: The previous chapter of the book described the processes for data analytics capability assessment and ranking to help organisations improve long-term value of their data analytics investments. This chapter will cover a prudential risk management case study from the financial services industry.

Prudential risk management

The objective of this case study is to demonstrate the importance of data analytics governance and capability that influence and impact value. The case study will also illustrate the application of data analytics in the area of Enterprise Support Services discussed in Chapter 8, and the use of 'tangible value', and 'notional value' covered in Chapter 16.

Brief background on prudential risk management

Commercial banks are generally very large with financial intermediation as their core business model that operates on the fundamentals of raising finance through deposit taking from customers, wholesale funding from the market through bonds, and from shareholder capital, which are used for lending. Depending on local regulations, many financial institutions offer other products and services, such as investments, superannuation, and insurance. In financial services institutions, the Risk Office is responsible and accountable for all aspects of risks, and Treasury in Finance is responsible for managing funding through an internal transfer pricing framework to facilitate risk transfer, business unit alignment, profitability measurement, capital allocation, and liquidity risk.

In this business model, the financial institutions encounter various types of risks. Credit risk arises due to the risk of default. Market risk arises due to a range of factors in the macroeconomy that affect foreign exchange rates, interest rates, inflation, and so on, which impact net interest income and net interest margin of the financial institutions. There are operational risks that arise due to factors such as early termination, and rollover of term deposits, mortgage prepayments, non-performing loan exposure, and hedging that impact cash flows and liquidity. In other words, due to the dynamic nature of behavioural cash flow and liquidity exposures, as well as contractual obligations on Options, the assets and liabilities in the balance sheet are continually in flux that can affect the stability of the financial institutions in balancing the intermediation process that is prone to liquidity exposures. Hence, to safeguard the interests of the depositors and/or policyholders of financial institutions, prudential regulations exist in each country that enforce compliance on the financial institutions. APRA is the Australian Prudential Regulation Authority.

APRA implements the international Basel standards, which are recommendations from the Basel Committee on Banking Supervision, through its prudential standards for Australian banks. The aim of the standards is to meet the financial services institutions' obligations to depositors and/or policyholders by requiring them to set up regulatory minimum capital requirements in order to safeguard the depositors and/or policyholders. The regulation provides two methods of determining minimum capital requirements, viz. **Standardised approach**, and **Internal Ratings Based (IRB) approach** respectively. IRB is now becoming the norm in the industry which is the focus of this case study.

The standards address different types of risks and includes credit risk, market risk and operational risk, and cover a range of metrics to measure compliance to the regulation. These include capital adequacy ratio, liquidity coverage ratio, net stable funding ratio, and value at risk (VaR).

This case study makes references to the publications of APRA for key requirements on Prudential Standard for Risk Management, Governance, Regulatory Compliance and Reporting, and related guides. All references to the relevant excerpts from the publications of APRA[44] are included within quotes in italics in the case study illustrations.

Overview of prudential risk management standard

APRA CPS 220 Prudential Standard Risk Management defines the objectives, and key requirements of Prudential Standard as:

'This Prudential Standard requires an APRA-regulated institution and a Head of a group to have systems for identifying, measuring, evaluating, monitoring, reporting, and controlling or mitigating material risks that may affect its ability, or the ability of the group it heads, to meet its obligations to depositors and/or policyholders. These systems, together with the structures, policies, processes and people supporting them, comprise an institution's or group's risk management framework'.

APRA CPG 220 is a guide that contains guidance on:

'Risk Governance, Risk management culture, Integration of the risk management framework

44 Australian Prudential Regulation Authority

and Internal Capital Adequacy Assessment Process, Risk management strategy, Risk management function, Compliance function, Monitoring and reporting'.

It also covers:

'*Scenario analyses: Stress testing, including both scenario analysis and sensitivity analysis is used to assess a range of potential impacts as a result of different material risks*'.

APRA CPG 235 Managing Data Risk contains guidance on:

'*The management of data and associated risks is important for a broad range of business objectives including meeting financial and other obligations to stakeholders, effective management and proper governance*'.

It covers:

'*Data life-cycle management, Data validation (Assessment of fitness for use, Data quality metrics), Monitoring and management of data issues, Data risk management assurance*'.

APRA CPG229 is related to:

'*The financial risks of climate change, Scenario analysis*'.

At its core, **APRA CPS 230 Cyber Risk** is:

' ... *about ensuring financial institutions remain operational during and after a cyber event. This requires layered security strategies that go beyond basic protections*'.

It includes:

'*Cross-functional involvement: Include teams beyond IT, such as legal, PR, and customer service in recovery drills*' and '*Resilience strategies: Redundant systems: Build backup systems to ensure uninterrupted service*'.

Introduction to the risk management case study

The focus of this case study is to illustrate how advanced data analytics for risk management can provide measurable value for a financial services institution. Additionally, it shows how data analytics value can be inhibited due to poor data management practices, and the importance of establishing data analytics capability ranking as a means of advancing to improved data management and governance, and value as a result. Also, this case study illustrates the use of 'tangible value', and 'notional value' in determining measurable benefits of advanced data analytics.

This case study references a publication by APRA[45] that discusses the benefits of the IRB approach to minimum capital requirements in, '*Is the capital benefit of being an advanced modelling bank justified?*'.

45 https://www.apra.gov.au/news-and-publications/capital-benefit-of-being-an-advanced-modelling-bank-justified

The publications says:

'Under the standardised approach, banks must use a common set of risk-weights prescribed by APRA in the risk-weighted assets calculation ... Standardised risk-weights are calibrated at a conservative level because they are less precise and apply to a wide range of banks ...While risk-weights are generally more conservative, there is a lower burden on standardised banks in terms of other supervisory requirements including data reporting'.

'Under the IRB approach, banks are permitted to use their internal risk models to produce estimates of credit risk such as the likelihood of borrowers not repaying their loans, and what loss that might cost the bank. The risk estimates are then input to risk-weight formulas prescribed by APRA to calculate credit risk-weighted assets ... The risk-weights under the IRB approach are tailored to the risks of an individual bank and are more precise than standardised risk-weights'.

'To use the IRB approach, banks must have very good historical data, a sophisticated risk measurement framework, develop advanced internal modelling capabilities, and undergo a rigorous assessment process in order to be accredited by APRA. IRB banks are subject to more stringent regulatory requirements, and more intensive ongoing supervision than standardised banks.'

'In principle, ***capital requirements under the IRB approach can be higher or lower than the standardised approach depending on the underlying risk characteristics of a specific exposure. On average, the capital framework has been calibrated such that IRB capital requirements tend to be lower than standardised capital requirements.'***

Data analytics capability ranking for prudential risk management

The highlighted texts above in the APRA publication suggest the following:

'On average, the capital framework has been calibrated such that IRB capital requirements tend to be lower than standardised capital requirements...' suggests that the business value of IRB approach to a financial institution is lower capital requirements which implies cost savings, and cost avoidance by optimising funding than the standardised approach. However, to realise these business benefits banks must,

- ***'have very good historical data'***
- ***have 'a sophisticated risk measurement framework, develop advanced internal modelling capabilities'***
- ***'undergo a rigorous assessment process in order to be accredited by APRA'.***

These requirements for advanced modelling capabilities translate to ensuring readiness of the financial institution for the following to embrace the IRB approach:

- Business and technology alignment to support rigorous assessment process
  - Workflows to support cross-functional effectiveness of, and compliance with, its internal control systems and business processes
  - Adaptability, ease of integration across risk types/categories, and with changing risk regulations as well as emergence of new categories of risks (e.g. Liquidity, Climate Risk, Cyber risk): Ref APRA CPG229
  - Stable, and scalable IT infrastructure ecosystem to support the capabilities – Ref APRA CPS 230

- Data management governance
  - Data management (data is current, accurate, and complete such that data quality is adequate for timely and accurate risk analysis) – Ref APRA CPG220, CPG235

- Data science management, and governance
  - Scenario analysis, and sensitivity analysis, and assess a range of potential impacts as a result of different material risks – Ref APRA CPG220, CPG229
  - End-to-end robust risk modelling, model performance monitoring, model governance, and modzsk management.

Integrated balance sheet asset and liability risk management

The following provide brief descriptions of the problems faced by commercial banks and how they impact a bank's ability to efficiently manage regulatory risks. The benefits of data analytics capability ranking as a means of making improvements in data analytics management and governance are also described below.

Problem statements: The information systems of large commercial banks have become very complex as a result of organic growth, as well as mergers and acquisitions, which result in the banks acquiring 'technical debt'. Due to the high costs involved replacing the acquired systems, and integrating them into the bank's original systems, many banks tend to operate the information technologies and systems in silos that lead to poor internal business processes in data management practices that involve the use of financial spreadsheets, manual interventions, frequent corrections, adjustments, and limited auditability which is shown in Figure 50.

As a consequence, this results in data duplication and errors that cause problems in assurance and trustworthiness of data quality. They also impact downstream processing for advanced analytics, and prediction of regulatory risk metrics, such as credit condition, liquidity, and net stable funding that impacts the balance sheet asset and liability positions.

Banking Risk Management Process – Current State

Involves financial spreadsheets, inflexible manual interventions, adjustments with limited auditability

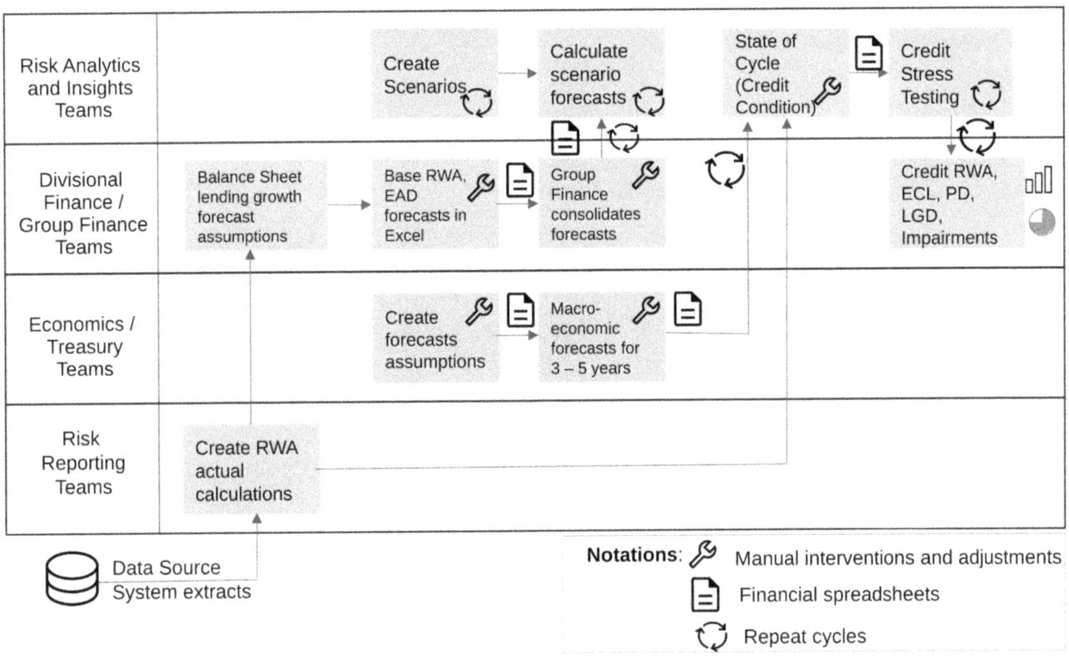

Figure 50: Banking Risk Management process – Current State

In referring to the example in Figure 50, teams from Risk Management, Finance and Treasury are involved in estimating, and preparing balance sheet lending forecasts, macro-economic forecasts, Risk Weighted Assets (RWA), Exposure at Default (EAD), Probability of Default (PD), Loss Given Default (LGD), and risk scenarios for stress testing the balance sheet. Due to the data management problems highlighted earlier, the internal processes involving manual interventions, financial spreadsheets, and frequent adjustments with repeated cycles results in limiting the number of risk scenarios for consideration, poor forecasting of risk exposures, and precise regulatory capital requirements as well as ineffective lending decisions.

Expected outcome of capability ranking: In recent times, large commercial banks have begun undertaking modernisation of their information systems driven by competition from leading FinTechs or digital banks (also known as neo-banks), who are replicating the BigTech business models by taking existing services, and transforming them into digital experiences that are now embedded in financial services, turning products into features to meet customer needs and keep the customers engaged.

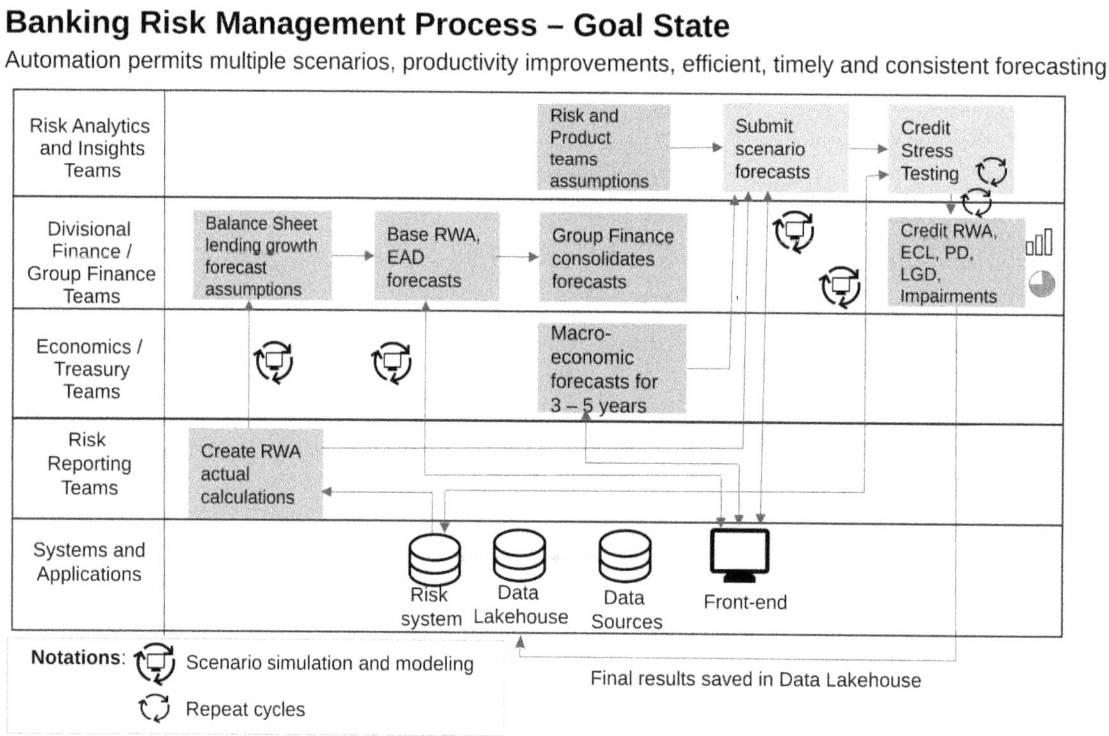

Figure 51: Banking Risk Management process – Goal State

As part of this digital transformation, large commercial banks have the opportunity to make improvements in their risk management practices by integrating and automating many of the tasks as shown Figure 51. In Figure 51, we can see that many of the manual interventions and tasks can now be eliminated by providing integration between systems and cross-functional teams that enable a consistent view of forecasts across the enterprise. System and team integration also help with automation and permits multiple risk scenarios to be simulated. It also improves productivity, efficiency, accuracy, and timeliness of forecasts.

Business benefits: By integrating and automating critical business processes as described in Figure 51, large commercial banks can reap the benefits of complex stress testing of balance sheet assets and liabilities, and meet enhanced regulatory compliance and risk reporting by using a consistent set of assumptions across different risk types. The banks will also be able to avoid erroneous lending and investment decisions. A large bank with $10 billion in annual revenue, and $400 billion in assets can expect to see the following estimated incremental benefits.

Greater governance and control in regulatory compliance: Capital add-ons is the normal mechanism that regulators impose on financial services organisations as a penalty for breach of compliance. Complex balance sheet stress testing and enhanced reporting avoids Regulatory Capital add-ons for the banks with the benefit in cost of capital savings. The use of 'notional value' is relevant for estimating the cost of capital savings associated with capital add-ons by considering the probability of occurrence of non-compliance for a bank, frequency of occurrence, as well as the amount of capital add-ons which can be obtained from historical trends in the industry.

Increased scope of scenario analyses: The bank will have the agility and ability to deliver ad-hoc stressed balance sheet scenarios and productivity, and efficiency in people and processes in reacting to change. The incremental benefits for this are tangible and can be computed by comparing the cost savings after the process changes described in Figure 51.

Greater efficiency: Incremental productivity benefits of $0.6 million due to cost reductions from efficiency and automation. Incremental revenue uplift of $1.16 million in better lending decisions. These benefits are tangible values that can be estimated due to better understanding of the portfolio and scenario modelling accuracy that lead to improved Risk Weighted Assets (RWA), Probability of Default (PD), and Exposure at Default (EAD) prediction accuracy, and net interest margin improvements.

Forecasting accuracy: Cost of capital incremental savings in regulatory capital adequacy forecasting of $1.8 million. Cash flow forecasting at product, portfolio and sub-unit levels can enable optimised liquidity management, asset, and liability management with potential incremental savings of $14 million. These benefits are tangible values that can be estimated from improved forecasting.

Integrated risk management capability ranking building blocks

Based on the current state and goal state risk management processes from Figure 50 and Figure 51, we can map the data analytics capability ranking for regulatory risk management as shown in Figure 52. It contains the capability dimensions by referencing APRA prudential regulatory risk management standards for the IRB approach. The capability ranking references the discussions in Chapter 21.

Figure 52 contains mapping of the characteristics of the capability dimensions against capability ranking, which provides the opportunity to initiate improvement processes in governance and management to achieve a higher ranking as goal state. The same principle can also be applied when a financial institution wants to benchmark its risk capability rank against its peers and/or competitors in the industry by using industry benchmarks[46] and publications by the regulator, and/or using published annual reports.

Capability Dimensions ↓ / Ranking →	Establish	Make-do	Govern	Adapt	Lead
Data management, governance - APRA CPG220, CPG235	Lack of metadata	High lead times to acquire and integrate data	Establish data dictionary	Self services across teams	Data Reuse and Data Ops
Workflows to support cross-functional process effectiveness	Error prone processes/manual documentation	Corrections, fixes and manual adjustments	Visualisation tools and integrated systems view	Model traceability and auditability	Automation of workflow tasks
Scenario analysis and sensitivity analyses – APRA CPG220, CPG229	Inability to run ad-hoc scenarios	Limited scenario run frequency	On demand execution of scenarios and forecasts		Repeatable cycles
End to end robust risk modelling, model governance and model risk management	No Quality Control of risk models	No reuse of prediction dependent variables; Models are hard to find and track		Automated progression across build / test / validation	Analytics Ops and Model Ops
Adaptability to changing risk regulations and ease of integration across risk types and categories APRA CPG229	Excel spreadsheets and manual adjustments	Lack of integrated view between risk types	Seamless integration and unified view across risk types and categories		
Scalable and stable IT infrastructure ecosystem – APRA CPS 230	Limited system resources and capacity	Time sharing of system resources across users	Scalable resources	Rapid system resource set up	Parallel processing and execution

Figure 52: Data analytics capability ranking for banking regulatory risk management

46 https://www.apra.gov.au/transforming-risk-culture-observations-from-apra%E2%80%99s-pilot-survey#cohort-benchmarking

CHAPTER 23

Illustrative case study of telecom industry

Overview: The previous chapter covered a financial services industry-focused prudential risk management case study. This chapter will describe a case study in telecom industry customer management with a potential for its extension using the evolving Agentic AI applications and related standards.

Telecom customer management

The objective of this case study is to illustrate the application of data analytics dealing with External value chain, and Internal value chain which we previously discussed in Chapter 8, with a focus on customer management. As part of this, the case study demonstrates the need for alignment of strategic management, business operations management, key performance indicators, and governance that influence and impact data analytics value contributions.

We will also discuss potential extension of the customer management case study by referring to the topics on autonomous decision-making, and Agentic AI workflow covered in Part VII – Human vs automation with AI – An interdisciplinary framework. The case study extension will also touch on evolving standards for multi-agent integration, and coordination such as Agent to Agent (A2A), and Model Context Protocol (MCP) by making references to the telecom industry Open Digital Architecture standards initiatives by TM Forum.[47] TM Forum is a global alliance of 800+ industry organisations across the connectivity ecosystem that includes top tier telecom infrastructure and service providers, network and software suppliers, and other industry related organisations.

Brief background on customer management

The telecom industry is a capital-intensive industry with multi-billions of dollars of investments required in communications network infrastructure to commission and operate vast varieties of fibre broadband and many concurrent generations of mobile network technologies ranging from 3G, 4G, 5G, and 6G. These technologies have been progressively deployed over many years to the spread capital spend. Due to mobility, along with continual transitions between 3G, 4G, and 5G cell towers, as well as indoors and outdoors WiFi networks, each of these technologies provide differing bandwidth and throughput capabilities that impact customers'

47 www.tmforum.org/oda

experience in using their mobile devices and apps.

The promise of 5G is huge with its potential to manage the exponential growth in demand for data across telecom networks, while presenting opportunities to supply even more data. This in turn is already enabling three overarching business applications:

Extreme mobile broadband supporting consumer and business use cases, such as virtual reality gaming and connected medical scanners

Critical machine-to-machine communication enabling reliable and low-latency applications, such as automated factories, smart mobility, and Internet of Things (IoT)

Massive machine-type communication supporting systems demanding high scalability and low power, such as smart cities, drone-based logistics, robotic process automation, and AI.

Overview of customer management

New technologies, including 5G and forthcoming 6G mobile networks, will accelerate digital transformation for telecom service providers by providing customers with an intuitive, seamless digital experience, which in turn will continue to make things easier for employees to manage customers. This includes new digital product launches that will allow telecom service providers to cease the sale of mobile plans and fibre broadband plans that have been managed with legacy systems. Tools like bill explainer are also key to providing digital services for thousands of customers by providing a clear and easy-to-understand bill, which is expected to reduce the number of billing-related complaints.

Mobile service providers face regulatory obligations to ensure network reliability like protecting critical infrastructure, notifying authorities of significant network changes, complying with emergency call service codes, and reporting on network performance. Telecom network operators and service providers, including mobile network service providers, are required to publish network reliability metrics and report on network performance to the country-specific regulators and authorities. Transition to 5G will also pave the way for improving customers' network experiences and reduce complaints to the industry ombudsman.

In order to cost-effectively manage customers, a 360° view of the customer is required across all touch points that includes accessibility, and connectivity to telecom networks, channels of interactions, customer service, product usage, and sensitivity to pricing.

Introduction to the customer management case study

The focus of this case study is to illustrate how advanced predictive modelling by leveraging customer journey events can provide measurable value for telecom service providers. Customers will not have to contact the telecom service provider if their needs can be anticipated and met with personalised solutions that arrive just when they are needed, which will provide a superior customer experience. This will require alignment of external value chain with internal value chain across the enterprise for effective customer management. It also underscores the need for sound data and analytics management practices to provide trustworthy customer information.

Problem statements: In order to meet regulatory obligations to deliver reliable network connection and acceptable network performance, as well as to see the benefits across a wide range of experiences (from more responsive gaming to new low-latency Agentic AI applications), the infrastructure will need to be closer to customers and nearby cell towers. And in order to process massive volumes of data, and manage superior customer experience, edge-computing infrastructure will be required. Multi-terabytes of data will need to be stored and analysed in real time in the same location where they are captured for the customer segments. Moreover, telecom service providers need to develop or source the software and connected device management capabilities needed to run advanced autonomous applications.

And consumers' willingness to pay for the services remains a challenge in a fiercely competitive industry with high rates of customer churn, and lower year-on-year revenue growth. To make the investments pay off, many service providers are on their journey towards a diversification strategy with a business-to-business-to-X business model and a focus on partnership with cloud providers and vertical industries with priority focus on utilities, transport and logistics, health care, and agriculture industries.

However, business operations in the new model must be aligned with strategic management, and will require optimising spend on customers, and be aligned with customer lifetime value, while also providing superior customer experience. Also, front-line teams, and autonomous applications will need access to more technologies to be integrated in the workflow in a service provider's new digital stack that can provide straight-through provisioning for orders, speeding up order time and improving the experience they can offer customers. But, as service providers continue on their digital transformation journeys, legacy systems, tools, and complex manual processes still remain.

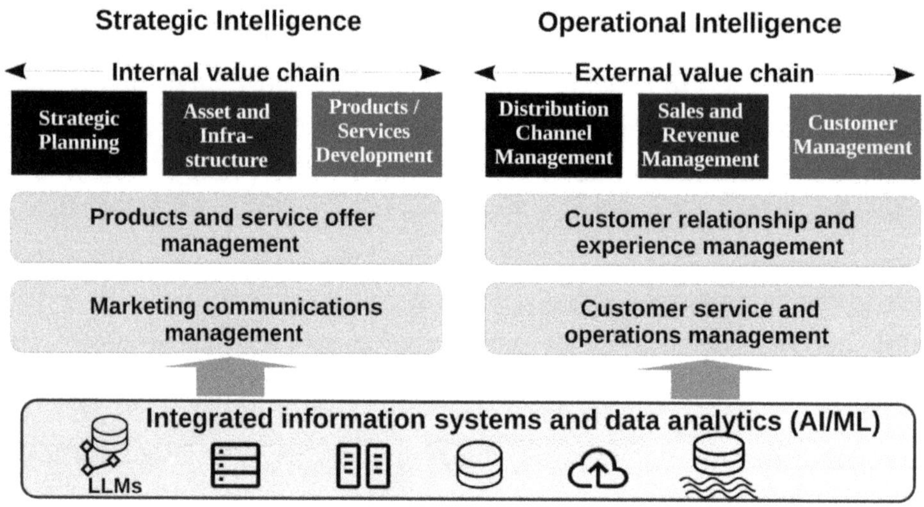

Figure 53: Aligning strategic and operational intelligence for customer management

How data analytics helps in effective customer management

Figure 53 represents a subset of Figure 18 in Chapter 8 that shows how organisational alignment and processes support effective customer management. Strategic intelligence involves market research, planning, policy development, governance, product, services and brand development, and marketing management support for front-line business units. Operational intelligence relates to customer relationship and experience management, as well as efficient operation of customer service and support across online and offline channels. There are many tasks involved in managing customers, many of which require human touch and involvement but which also require support by integrated information systems comprising front-office and back-office applications, along with data analytics that are also instrumental in autonomous decisioning in business processes and workflows.

The following provide brief descriptions of a number of use cases involved in managing customers, some of which deal with internal value chain and others with external value chain. These use cases consider customers' preferences and are subject to privacy protection policies.

Customer Behaviour Analysis: This use case analyses and identifies customer needs and interests based on customer lifestyle attributes of product, service and communication network usage, as well as channel interactions. Such analyses help to develop customer journey views that are used in predictive and preventive models, segmentation and targeting effectiveness for next-best action, and proactive engagement with customers.

Customer Journey Analytics: This involves predictive modelling by integrating various customer events including interactions across multi-channel customer engagements. The predictive models help to determine propensity of customer experience, dissatisfaction, complaints, and potential churn. Markov Chain together with Sankey diagram visualisation helps in developing transition matrix for predicting next possible event which can be used for various next best actions related to churn prevention, complaint reduction, improved customer experience, and so on. Figure 54 illustrates the journey event steps with an example of a customer's network experience, call centre, and retail outlet experience that led to customer complaint and churn. Thickness of the lines indicates density of events.

These types of models have been useful in proactive actions for tariff optimisation and customer retention by understanding customer's sensitivity to pricing based on product/service/network usage, and improvement of customer experience. Reduced churn adds to the bottom line in financial statements with gain of retained revenue and reduced re-acquisition costs.

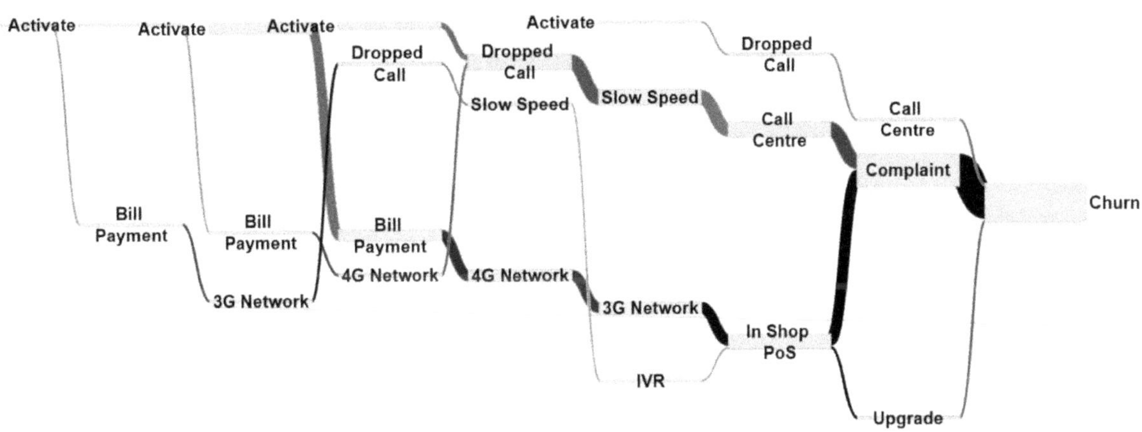

Figure 54: Customer journey event steps leading to complaint and churn

Customer Network Experience Analytics: Telecom network operations teams take advantage of massive volumes of control plane, and user plane signalling data generated by mobile networks representing customers' connection, access, and usage. These are used for multiple purposes by different business units (i.e. Finance, Marketing, Network operation) in telecom as shown in Figure 48. One of the usages of such data is also for representing customers' mobility across different networks by applying geo-spatial analytics, as well as customers' journey steps, when interacting with the telecom service provider as multiple experiences has shown in Figure 54. They are also useful for tracking regulatory compliance, and quality of customers' perceptual experience by applying advanced analytics including geo-spatial analytics to identify network congestion around cell tower locations.

Analytics also includes connected device performance to isolate network performance issues caused by end-user devices and applications. Some of the devices and applications on the devices are generally outdated, so they could malfunction and affect the network. These analytics also enable corporate account teams and Customer Care to manage corporate customers more effectively. Through targeted and proactive network performance improvement activities, corporate customer satisfaction is improved.

Optimise Mobile Radio Access Network Assets: Mobile radio network sites are characterised by dynamic traffic patterns caused by mass movement and congregations of people at varying times of the day. To avoid performance issues and improve customer experience, the network operations team analyses and optimises the cell tower access networks and core networks. Figure 55 illustrates this point with visual representation of mobile device usage by customers along commuter corridors in Melbourne, Australia. It shows the impact on mobile networks due to communication signalling when hundreds and thousands of customers simultaneously arrive in Melbourne city centre, and other nearby geographical locations, from different suburban locations, and during morning peak hour traffic. This pattern varies throughout the day with more diverse patterns during lunch time and evening peak hours. Thickness of the lines indicates density of mobile device connections.

Value-based Network Capacity Planning: Optimising telecom network also involves planning for network geographical expansion (i.e. deploy new cell towers), and capacity increase (i.e. upgrade cell towers) to relieve network congestion and to improve customers network experience. Leveraging network and customer data helps to make better informed, and accurate targeted decisions on network investments. Geographical expansion involves forward planning, forecasting, and large upfront capital spend with unpredictable revenues. Predictive analytics helps with better utilisation of capital spend for new cell site investments based on customer mobility and usage trends in target locations rather than relying on population growth alone. It improves revenue from cell sites that have corresponding usage growth, increases customer satisfaction, and means a better customer lifetime value.

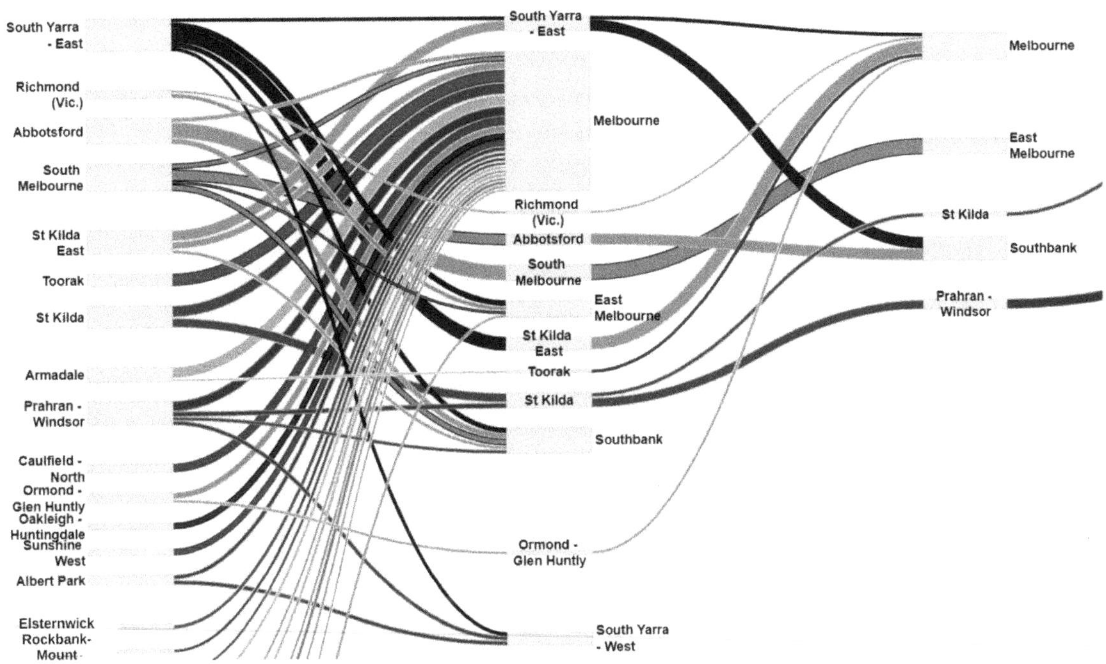

Figure 55: Mobile device usage along commuter corridors during morning peak hour

Sentiment Analysis: Feedback from customers across all touch points is extremely useful in order to proactively take action on negative events and identify topics related to negative feedback. Advanced text analytics using natural language processing (NLP), Latent Dirichlet Allocation (LDA), cosine similarity, and sentiment analysis using part of speech tagging not only helps to determine perceptual experiences of customers, and how they are likely to share their frustration and satisfaction, as well as in determining next best action. The advanced analytics helps in formulating forward planning and strategy development by analysing grouping of topics and similarity of sentiments.

With increasing availability of large language models, Agentic AI applications can take advantage of LLMs for sentiment analyses. However, the economics of using LLMs versus the advanced analytics techniques mentioned earlier will need to be evaluated.

Customer Experience Index: The customer experience index provides telecom service providers with a true measure that takes all 'touch points' into account and provides insights into how customers' perceptual experience of products, services, and network rather than simply using quality of service (QoS) metrics. The index includes both direct and indirect data and is based on multiple experiences and interactions of individual customers as shown in Figure 54.

Improvement in Risk Scoring Models: Customer experience index helps to predict the risk of complaints that in turn assists in taking steps to actively reduce the complaint rate among the customer base or key customer segments. Proactively addressing customer service issues and complaints early are also instrumental in reducing the cost of processing and rectifying complaints, and more importantly minimises the cost of complaint resolution and complaint-related customer churn. Further improvements in risk management can be made by using various behavioural and predictive models. Besides reducing complaints and churn, the models help reduce regulatory risks and fines.

Customer Satisfaction/(NPS): Sentiment analyses, customer experience analyses, and risk models can all be used to identify the root causes of low customer satisfaction as well as new, emerging trends of issues. Together with customer experience index, net promoter score (NPS) is used to measure customers' level of satisfaction in using products, services, and network.

Customer Lifetime Value: Lifetime value of customer is an important measure for ensuring customer-level profitability that determines effective ways of managing customer interactions with the telecom service providers. This can be achieved by providing appropriate treatments and responses for customer events, whether the events are generated during human interaction or machine generated, such as telecom network usage.

Next Best Action: Together with sentiment analyses, customer experience index, risk scoring models, customer satisfaction, and customer lifetime value, the next best action for the customer is taken following an event that requires attention of the service provider. Telecom service providers also periodically perform proactive next best action for the customer using segmentation and targeting models which helps to improve customer experience and satisfaction that lead to lasting customer lifetime value.

Customer Service Empowerment: In order to quickly identify/diagnose/react to events and incidents that are likely to impact customers, the provision of 360° view of real-time events, the true customer experience index and related KPIs to Customer Service agents and front-line staff are important as part of the internal value chain. With the advent of Agentic AI applications, many such events can also be rapidly analysed and resolved with autonomous decisioning that can lessen the burden on front-line staff, thereby optimising costs while also improving customer experience and lifetime value.

Expected outcome:

Figure 56 provides a high-level architecture representing realisation of the customer management use cases by integrating data analytics in business processes. Automated decisioning in business process steps can be predictive models (i.e. inductive reasoning) and/or rules engine (i.e. deductive reasoning) as previously covered in Chapter 13, 'Automation and autonomous decision-making with AI'. The Integration Services layer can involve implementation using a number of different technologies, with most common being API integration (e.g. REST API). With the emergence of Model Context Protocol, and Agent-to-Agent protocol, MCP and A2A are also candidates for conversational Agentic AI application integration (e.g. OpenAI) along with LLMs and RAG.

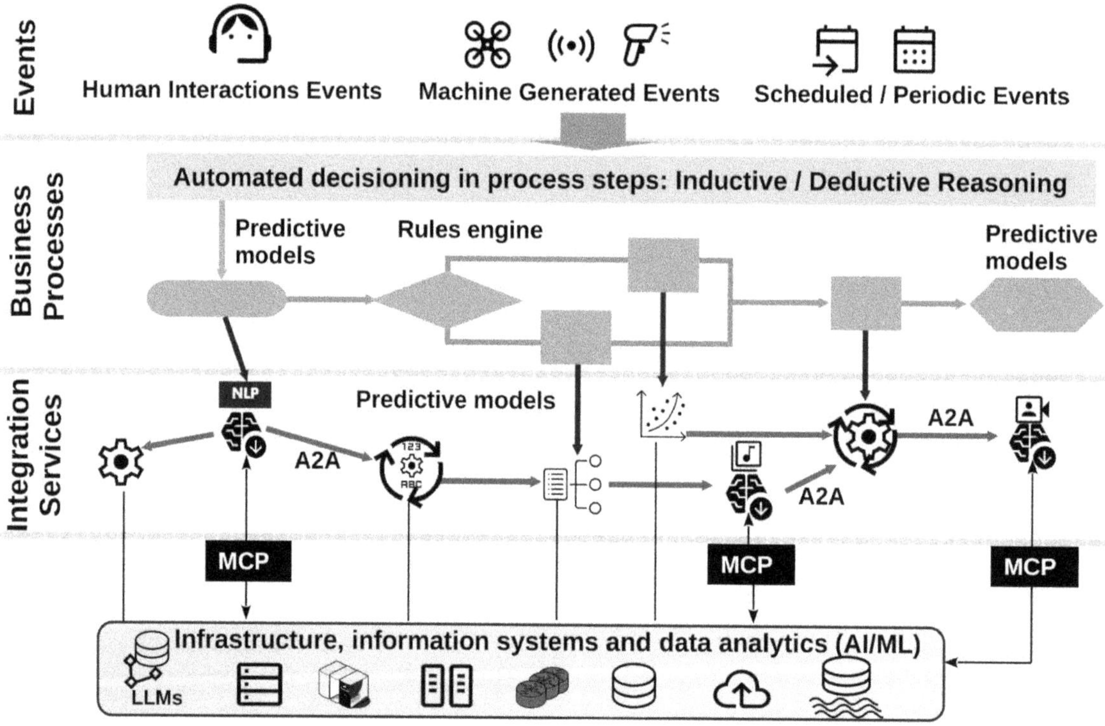

MCP: Model Context Protocol; **A2A**: Agent to Agent Protocol

Figure 56: Autonomous decision making in customer management

By aligning data analytics investments with strategic and operational management processes, telecom service providers can expect positive financial benefits for effective and efficient customer management. Tables 48 to 52 describe a range of benefits enabled by data analytics along with aspects of data analytics that are attributable to quantified business value.

Table 48 contains benefits of integrated data for managing customers and channels effectively.

Table 48: Benefits of integrated data for customer management

Customer Management Benefit Areas	Benefits Enabled by Data Analytics	Benefits Attributable to Data Analytics
Benefits of Integrated Data		
Improved insights.	Better predictive models, faster to deploy and economical to run.	Improved availability and breadth of variables allow widening coverage of target customers per segment; Successful customer contact and response rates.
Ease of access to information to enable decision-making.	Reduced time to decision.	Improved availability of contact history and holistic customer information (e.g. profitability, product usage, billing complaints, network/service experience etc.).
Improved quality of decisions from facts.		
Improved operational efficiencies.	Reduced overhead.	Improved reliability and accuracy of holistic customer information.
Reduced infrastructure costs in support / maintenance.	Reduced operating expenses.	Consolidated and integrated data.

Table 49 and Table 50 contain benefits of data analytics for product and marketing teams to support front-line teams. These tables illustrate the value of aligning internal and external value chains in the enterprise.

Table 49: Benefits of data analytics for product, and marketing teams to support front-line teams

Customer Management Benefit Areas	Benefits Enabled by Data Analytics	Benefits Attributable to Data Analytics
Integrated View of Customers and Channels		
A. Segmentation Models		
Effectiveness of marketing strategies and competitiveness.	Improved visibility to product/proposition performance and customer profitability.	Improved visibility of customer value, offer, usage and contact information, product proposition, pricing and history;
Effectiveness of product development.		
Improved customer loyalty.	Increased rate of loyal customers.	Improved availability of customer value, tenure and usage history.
B. Targeting Effectiveness		
Perform complex modelling with enriched data for pricing, churn, propensity and attribution models.	Increased agility; Better predictive models, faster to deploy and economical to run; Reduced time, resources and costs for segmenting and targeting.	Improved availability and breadth of variables allow widening coverage of target customers per segment; Successful response rates.
Better segmentation.	Improved unique customers per segment; Improved cost to revenue ratio per segment.	Improved unique customers in market segments; Aggregated view of multiple services / accounts.
Increased accuracy of targeting.	Shift in multi-channel / offer type mix; Increased revenue; Customer satisfaction through relevancy of contact.	Improved latency, quality and breadth of information.

Table 50: Benefits of data analytics for front-line customer service teams, and channels

Customer Management Benefit Areas	Benefits Enabled by Data Analytics	Benefits Attributable to Data Analytics
Integrated View of Customers and Channels		
C. Contact Effectiveness		
Increased response rate.	Improved multi-channel and mix of contacts consisting of non-targeted, targeted product and service-centric, event-based across appropriate channels.	Improved reliability and accuracy of holistic customer information; Improved visibility of customer value, offer and contact information.
Increased revenue.	Increased revenue per contact.	
Increased customer retention.	Customer satisfaction through relevancy of contact; Increased retained revenue.	
D. Customer Contact Cost Reduction		
Integrity of customer data - eliminate duplication and redundancy.	Reduced cost per contact; Improved customer satisfaction through timely, relevant contacts.	Reduced redundancy of records; Aggregated view of multiple services / accounts; Monitoring of channel costs and revenue.
Reduced service delivery costs across right channel.	Reduced average cost per contact; Optimum channel mix.	

Table 51 contains benefits of data analytics for front-line customer service teams and channels.

Table 51: Benefits of data analytics for front-line customer service teams, and channels

<table>
<tr><th>Customer Management Benefit Areas</th><th>Benefits Enabled by Data Analytics</th><th>Benefits Attributable to Data Analytics</th></tr>
<tr><td>Integrated View of Customers and Channels</td><td></td><td></td></tr>
<tr><td>Customer Service</td><td></td><td></td></tr>
<tr><td>Cross-channel coordination for Next Best Activity (NBA).</td><td>Improved multi-channel and mix of contacts consisting of non-targeted, targeted product and service-centric, event-based across appropriate channels.</td><td>Improved reliability and accuracy of holistic customer information.</td></tr>
<tr><td>Efficiency and productivity of customer service agents.</td><td>Improved cost to revenue ratio per agent.</td><td rowspan="4">Improved availability of contact history and holistic customer information (e.g. profitability, usage, billing complaints, service experience etc).</td></tr>
<tr><td>Reduced call handling time due to better access to information.</td><td>Improved rate of first-call resolution of problems.</td></tr>
<tr><td>Optimised call routing and handling.</td><td>Reduced rate of IVR call diversion to agent.</td></tr>
<tr><td>Improved performance tracking.</td><td>Improved cost to revenue ratio per agent.</td></tr>
<tr><td>Increased effectiveness of NBA.</td><td>Increased revenue per contact.</td><td>Improved reliability and accuracy of holistic customer information.</td></tr>
<tr><td>Improved customer satisfaction.</td><td>Increased retained revenue.</td><td>Improved visibility of customer value, offer and contact information.</td></tr>
</table>

Table 52 contains benefits of data analytics in enriching customer value as well as financial performance of the organisation.

Table 52: Benefits of data analytics in enriching customer value and financial performance of the organisation

Customer Management Benefit Areas	Benefits Enabled by Data Analytics	Benefits Attributable to Data Analytics
Integrated View of Customers and Channels		
Customer Value Enrichment		
Optimised cost to retain (CTR) / cost to acquire (CTA) based on customer value.	Reduced costs for CTR/CTA.	Improved availability of and access to customer value and customer history; Aggregated view of multiple services / accounts.
Channel optimisation based on customer value.	Shift in Channel / Offer type mix; Increased revenue; Improved channel profitability.	
Optimise service costs based on customer value.	Reduced average cost of service.	
Optimise pricing based on customer value.	Improved product take up rate and profitability.	Improved availability of customer value, tenure, usage history and proposition pricing.
Improved accuracy on profitability.	Improved revenue and cost per customer.	Improved availability of customer value and direct costs (e.g. Revenue and discount by product, usage etc).
Improved cash flow.	Improved borrowing costs, debt/revenue ratio.	Improved availability of customer revenue, budgeting and forecasting information.
Financial Performance Management, Budgeting and Forecasting		
Consistency and alignment of forecasts.	Improved accuracy and reduced cost.	Improved availability of sales channel performance and holistic customer revenue, budgeting and forecasting information.
Reduced overhead by streamlined forecasting.	Reduced cost.	
Improved performance management of channels / partners.	Improved cost:revenue ratio per channel / partners.	

Business benefits: By integrating and automating critical business processes for customer management, large telecom networks and service providers can reap the benefits in a capital intensive and hyper-competitive industry. A large telecom service provider with $10 billion in annual revenue with a year-on-year revenue growth rate of 5% can expect to see the following annual incremental benefits in data analytics value as a result of effective and efficient customer management.

Internal value chain related:
Data analytics value attributable to customer churn reduction and retained revenue improvements of $10.2 million per annum.

Data analytics value attributable to reduction in number of contacts per customer and associated reduction in costs; increased offer take-up rate due to effective segmentation and targeting, and associated increase in revenue uplift of $11.4 million per annum.

External value chain related:
Data analytics value attributable to customer service cost savings in automated call diversion, reduction of repeat calls, reduced cost per call, reduction in problem diagnostics and resolution time, and associated cost reduction of $5.4m per annum.

At the time of writing this book the number of successful case studies that took advantage of Agentic AI applications with proven ROI is far and few due to a range of problems, including limitations of these technologies to integrate with organisations' business processes and workflows, evolving standards such as Model Context Protocol (MCP) and Agent to Agent (A2A) protocol, Knowledge Graph, ontology, and high costs associated with deploying the technologies on large scale.

However, these technologies are expected to provide 20% to 30% improvements to data analytics value for both internal and external value chains when AI is focused on narrow domain specific applications rather than general purpose conversational AI. The next chapter illustrates, with a case study, how Agentic AI systems focused on domain specific applications are becoming capable of performing routine cognitive tasks involving digital technologies that are performed by experts across administrative, financial, and professional services.

CHAPTER 24

Case study of Agentic AI use in financial services industry

Overview: Agentic AI's new capabilities in LLM and NLP, such as chain-of-thought, reasoning, MCP, A2A, are enabling it to autonomously produce digital deliverables of routine cognitive tasks involving such examples as the ones below.[48]

- Auditing tasks that require creation of spreadsheet to complete calculation of sample size for audit testing with 90% confidence level and a 10% tolerable error rate. Performance of variance analysis on two consecutive periods of financial data and reporting these in the spreadsheet for review by executive team for their critical decision-making and overall financial oversight.
- Vendor sourcing in supply chain management involving creation of a short document outlining a revised sourcing and nomination workflow that includes a 'Modular quotation structure' to develop a plug-and-play model for cost drivers based on features, design elements, and bill of materials used.

This chapter develops a business case for Agentic AI systems for anti-money laundering (AML) in financial services industry with a banking example.

Overview of anti-money laundering regulation

AUSTRAC[49] (Australian Transaction Reports and Analysis Centre) defines its role as below:

"AUSTRAC performs a dual role as Australia's anti-money laundering and counter-terrorism financing (AML/CTF) regulator and financial intelligence unit. This dual role helps to build resilience in the financial system and ***enables AUSTRAC to use financial intelligence and regulation to disrupt money laundering, terrorism financing and other serious crime****.*

As Australia's AML/CTF regulator, we regulate more than 19,000 businesses that provide financial, gambling, bullion, remittance and digital currency exchange services. We ensure regulated

48 https://openai.com/index/gdpval/

49 https://www.austrac.gov.au/business/core-guidance/reporting

businesses comply with their obligations to have systems and controls in place to manage their risks and protect them and the community from criminal abuse.

As a financial intelligence unit, we collect and analyse financial reports and information to generate financial intelligence *that contributes to law enforcement and national security investigations.* ***Our specialist analysts generate targeted, actionable intelligence and work closely with industry, government and law enforcement partners to deliver tangible investigative and operational outcomes.***"

As part of the regulation for reporting obligation, AUSTRAC requires regulated entities such as the financial service organisations (e.g. banks, insurance companies, currency traders) to submit the following:

Threshold transaction reports (TTR) for transfers of A$10,000 or more in cash (or the foreign currency equivalent). TTRs are due within 10 business days after the date of the transaction.

International funds transfer instruction reports (IFTI) for transfers of funds of any value into or out of Australia, made either electronically or under a designated remittance arrangement. IFTIs are due within 10 business days after the transfer instruction is sent or received.

Suspicious matter reports (SMR) for a suspicion that a customer or transaction is related to criminal activity. This must be submitted within 24 hours of becoming suspicious if it relates to terrorism financing, or within three business days for anything else.

Cross-border movements of physical currency of A$10,000 (or the foreign currency equivalent) or more if carrying by mail or shipping money into or out of Australia.
In addition, a financial services organisation is required to submit compliance reports to AUSTRAC, when required by AUSTRAC, that explain how the organisation is complying with the AML/CTF Act, Regulations and AML/CTF Rules.

AUSTRAC coordinates with a wide range of domestic partner agencies and international counterparts and **periodically collects, publishes and shares typologies[50] and case studies reports.** These are intended to inform industry and the wider community about the various methods criminals use to conceal, launder or move illicit funds and to commit financial or other crimes. **This information assists industry to strengthen its measures to detect money laundering activity** and protect both businesses and customers from criminal activity.

50 https://www.austrac.gov.au/search?keys=typology

Introduction to the anti-money laundering case study

Commercial banks encounter vulnerabilities associated with their customers' transactions with the bank that sometimes conceal suspicious activities. AUSTRAC requires banks to review regulatory requirements and typologies to:

- determine what money laundering and terrorism financing vulnerabilities are most relevant to their organisation
- guide training programs they provide to staff to raise awareness of the issues within the organisation
- assess risks
- assist them in identifying and investigating unusual customer activity
- use the risk 'indicators' as a guide when describing unusual behaviour in suspicious matter reports
- refine existing and/or add new detection scenarios and rules they use in their transaction monitoring programs and
- highlight the benefits of maintaining a robust AML/CTF regime within their organisation.

The focus of this case study is to develop a business case justifying the financial benefits of implementing advanced data analytics involving the use of Agentic AI systems.

Agentic AI for anti-money laundering regulatory reporting

The following provides a brief description of the problems faced by commercial banks, how they impact the banks' abilities to efficiently manage regulatory risks. The benefits of data analytics with Agentic AI as a means of complying with regulations is also described below.

Problem statements: Large commercial banks have several million customers who perform periodic transactions with their bank. The volume and types of transactions handled by banks are also complex and huge that run into several million per day. In accordance with AUSTRAC's regulation, the banks are required to monitor these high volumes of transactions and the banks' customer activities to detect and report abnormal TTR, IFTI, or SMR transactions to AUSTRAC within the required times indicated by the regulator.

Banks have deployed data analytics to deal with such requirements for many years along with several manual tasks performed by business analysts and investigators of the bank.

However, the rate of increase in volume and complexities of transactions as well as the various methods criminals use to conceal, launder or move illicit funds and to commit financial or other crimes are also on the increase as reported by industry analysts.[51]

Banks' analysts spend enormous amounts of time investigating potential financial crimes, known as *alerts*, with tedious, manual analysis of unstructured data to identify relevant entities and individuals. Many of these alerts can result in *false positives* implying false alarms that consume wasteful human resources dedicated to such investigations.

For instance, a large commercial bank with 20 million customers generates about 2,500 suspicious transaction/matter alerts every day. It usually takes about 60 minutes on average for manual investigation of each of these alerts; that requires a head count of 300 business analysts and investigators which can cost the bank about $30 million. Many of the alerts usually end up being *false positives* at a rate of about 85% which translates to about $25 million of wasteful investigations. As a result, most banks around the world focus on improving the quality of alerts and efficiency as a result.

Also, creating quality reports to regulators has its share of difficulties due to additional time spent on them. There is also opportunity cost in delayed reporting that can make it harder to prevent further crime.

While many banks concentrate on efficiency gains by reducing *false positives,* they often ignore *false negatives* which cause more serious harm and damage to the banks. *False negatives* are '*Trojan horses*' that go undetected with serious implications for failed reporting of suspicious transactions/matters to regulators.

These problems pose challenges for the banks to keep up with such trends to accurately report suspicious activities to regulators. Failing to report these has severe consequences to the banks that involve 'Enforcement actions' with specific legal actions such as:

- civil penalty orders
- enforceable undertakings
- infringement notices and
- remedial directions.

51 https://www.ey.com/en_se/insights/financial-services/how-ai-is-reshaping-the-future-of-transaction-monitoring

Regulators have also imposed penalties and fines to banks to the tune of several hundreds of millions of dollars[52] which also causes reputation damage to the organisation.

Agentic AI solution overview: Banks have employed data analytics for many years with advanced machine learning capabilities for detecting suspicious transactions/matters to the regulator. However, due to the complexities involved in the layered banking transactions that hide their activities, limitations of the banks IT systems and the need for analysing unstructured information in documents, banks spend enormous amounts of effort in manual investigations and reporting.

With the recent advancements in Agentic AI research, development and trials, it offers the following short list of capabilities that are relevant for anti-money laundering applications.

- Extraction and processing of unstructured data from documents and presentation of facts in summarised form using NLP
- Integration of entity/individual's specific customer profile and accounts across different IT systems with LLMs to create context for suspicious matter investigations
- Named entity identification of related parties in layered transactions to detect individuals and locations of suspicious activities. This is specifically relevant for Trade Finance transactions that are often very complex with cross-border involvement spanning different geographies
- Advanced machine learning algorithm integration using Support Vector Machines and Distribution, KNN (k nearest neighbour) and anomaly detection to identify *false negatives*
- Cosine similarity and topic modelling (i.e. Latent Dirichlet Allocation) to analyse suspicious activities with AUSTRAC publication of typologies to detect and classify *alerts* for priority action
- Multi-agent coordination with automated workflows with human oversight when required
- Consistent and standardised report generation in accordance with the requirements of regulators and
- Sophisticated prompt engineering with precise instructions and reference materials supplied to AI agents for automation of workflows and autonomous decision-making.

52 https://www.austrac.gov.au/business/core-guidance/consequences-not-complying and https://www.asic.gov.au/about-asic/asic-investigations-and-enforcement/fines-and-penalties/

Expected business benefits: By integrating and automating the critical suspicious activities investigation processes described above, large commercial banks can benefit from efficiencies in alert investigation and regulatory reporting. A modest reduction in *false positives* can result in savings of $5 million to $10 million per year. In addition, with advanced machine learning algorithms to detect *false negatives* quicker, a bank can avoid regulatory fines and reputation damages to the tune of several hundred millions of dollars in financial losses.

Part XI

Concluding Thoughts

This part of the book summarises the previous parts, and chapters of this book with concluding thoughts.

CHAPTER 25

Redefining value where people and AI meet

We open this book with insights drawn from consulting practice and organisational research, highlighting a persistent challenge faced by organisations: accurately measuring the true returns on data analytics investments. While data analytics initiatives consistently generate value, the difficulty lies in how this value is identified, articulated, and communicated across the enterprise. These challenges arise from multiple factors, including limited recognition of data analytics investments as organisational assets within audited accounting frameworks, as well as the absence of mature data analytics management and value governance practices.

A similar challenge exists at the individual person level within organisations. However, unlike data analytics, human capital is supported by well-established governance frameworks encompassing talent acquisition, performance management, and incentive systems. This contrast underscores a structural gap in how data analytics value is governed and institutionalised.

Building on these observations, this book proposes alternative approaches to measuring, attributing, and managing data analytics value. Central to this perspective is the treatment of data analytics as a long-term, renewable, and growth-enabling enterprise asset, rather than as a short-lived or consumable resource. In doing so, the book moves beyond simplistic analogies that frame data as an extractive commodity such as 'crude oil' and instead advances a more sustainable and strategic view of data analytics as a renewable source of enduring organisational value.

Digital economy and organisational value chain

Much like other foundational enablers of modern economies, data analytics now sits at the centre of organisational business processes that underpin the digital economy. However, a critical distinction must be made between the broader notion of a 'data economy' and the economic value of data analytics within an individual organisation. At the enterprise level, value is realised not through data or analytics in isolation, but through their alignment with appropriate theories of value, organisational strategy, objectives, and performance metrics. Such alignment is essential to ensure that data analytics investments translate into measurable business outcome across the organisation's value chain, including processes within and across business functions, sub-functions, and job roles.

These processes ultimately depend on the activities and tasks performed by people. In most organisations, human resources represent the largest category of expenditure; for example, in a major financial services organisation, personnel costs account for approximately 61% of total operating expenses. Although this proportion varies by industry, people remain the dominant resource base in service-oriented and knowledge-intensive organisations. Human capital is therefore central to organisational performance, as it provides the unique sources of knowledge, judgement, intuition, and decision-making required to create value.

At the same time, research in psychology has long established the cognitive and memory limitations that constrain individuals' ability to perform complex tasks optimally in isolation. Consequently, employees rely on organisational resources, particularly IT to augment their capabilities and enable more efficient and effective task execution. These supporting resources include collaboration and knowledge-sharing tools, inter-functional communication infrastructure, and enterprise information systems such as customer relationship management (CRM), enterprise resource planning (ERP), and financial accounting and reporting systems. In the same financial services organisation, such secondary resources account for approximately 18% of total operating expenses. While data analytics expenditures are not typically reported as a distinct category, they are generally embedded within IT spending, further underscoring the need for clearer recognition and governance of analytics as a strategic organisational asset.

Value attribution in the age of AI

When senior leaders attempt to evaluate value contributions, often framed as ROI of an organisation's core resources, including people, technology, and other enabling assets, they are frequently confronted with a lack of clear and actionable insight. This challenge is evident where isolating and attributing the respective contributions of these resources proves difficult in practice. The issue is particularly acute for data analytics, whose impact is embedded across diverse job roles, work activities, and business functions, each with distinct operating contexts and performance drivers. The complexity is further heightened by the coexistence of renewable and depleting resources within the organisation's operating model, as well as by the rapid emergence of agentic AI applications that blur traditional boundaries between human and technological value creation.

In response to these challenges, this book takes a pragmatic step back to examine the evolution of IT and data analytics over the past five decades, focusing on how organisational perceptions of value, stakeholder expectations, and measurement practices have changed over time. This review highlights persistent gaps in existing approaches and offers practical recommendations for improvement. Drawing on real-world examples, the evidence shows that organisations that move beyond data silos and invest in data integration, reuse, and multi-genre analytics are better positioned to realise both immediate cost efficiencies and sustained value growth. These insights are intended to support senior management in making more informed investment, governance, and prioritisation decisions regarding data analytics as a strategic organisational capability.

Interdisciplinary framework

This book introduces a strategic framework for governing and optimising human and data analytics resources as renewable enterprise assets that underpin judgement, decision-making, and sustained business performance. The framework provides senior leaders with a structured approach to understanding where value is created, where risk is introduced, and how accountability should be assigned when decisions are supported or driven by analytics and AI. It explicitly addresses the complementary strengths and limitations of human judgement and data-driven models, including the biases inherent in both, and offers guidance on how these trade-offs should be managed at the enterprise level.

From a strategic perspective, the framework distinguishes between activities where automation and agentic AI can responsibly substitute human effort, primarily routine, low-risk tasks and those where analytics should augment human judgement, particularly in contexts involving uncertainty, inductive reasoning, and high-impact decisions. In areas of elevated risk or material value creation, the framework reinforces the importance of human-in-the-loop decision models to ensure appropriate oversight, accountability, and risk management.

To support execution, the framework is underpinned by practical governance and value-management tools, including DAVA, DAVE, and DAVO, alongside industry benchmarks, use cases, and case studies from the telecommunications and financial services sectors. It also introduces structured methods for assessing organisational readiness and levels in data analytics capabilities, enabling boards and executives to prioritise investments and monitor progress over time.

At its core, the framework positions data analytics value governance as an enterprise-level discipline, essential for protecting, growing, and realising the long-term value of data analytics investments as strategic, growth-enabling assets of the organisation.

Guiding principles for effective data analytics (AI/ML) value

The evolution of IT and data analytics has transformed the organisational landscape. Today's enterprises operate within an ecosystem saturated with vast volumes of data, increasingly sophisticated predictive models, large-scale language models, and rapidly advancing forms of

AI embedded in NLP, image recognition, audio translation, and agentic AI systems. Together, these capabilities are reshaping how work is performed and, in some cases, substituting elements of human intelligence.

Yet this abundance of analytical capability has not been matched by equivalent progress in how organisations define, govern, and realise value from intelligence, whether human or artificial. The complexity of the ecosystem, combined with legacy accounting, measurement, and governance practices, has made the true value of data analytics both difficult to see and easy to misrepresent. This book argues that **redefining value at the intersection of human and data analytics investments** is no longer optional; it is a strategic imperative.

As a closing synthesis, the following principles capture the core insights of this work and provide practical guidance for leaders navigating this transition:

Avoid the balance sheet black hole. Traditional financial statements fail to capture what matters most in modern enterprises: intelligence embodied in people and data analytics. Organisations must move beyond conventional metrics and adopt alternative approaches that recognise intelligence as a renewable, value-creating asset, whether human or AI-enabled.

Measure smarter. Value does not emerge from more data or more models, but from the ability to measure and attribute their impact on business outcome and communicate value effectively. The tools and frameworks presented in this book are designed to make data analytics value visible, comparable, and actionable for decision-makers.

Establish what intelligence is really worth. The strategic question is not whether to invest in human or artificial intelligence, but how to determine the value of each and allocate investment accordingly. This requires explicit consideration of risk, reward, accountability, and the context in which intelligence is applied.

In an ecosystem that continues to evolve at speed, progress requires focus. Organisations should begin by concentrating on their most material data analytics and AI resources, i.e. those with the highest levels of usage and potential impact, using pragmatic approximations of value to avoid analysis-paralysis. Industry benchmarks provide a valuable starting point, enabling organisations to assess investment effectiveness, evaluate the mix of resources embedded in roles and processes, and determine where automation is appropriate and where human judgement remains essential to value creation.

Finally, value measurement must be understood as a journey, not an endpoint. Establishing the value of intelligence is only the first step. Sustained impact depends on the development of strong data analytics value governance and management capabilities – capabilities that allow organisations to continually identify value enablers and inhibitors, refine their use of intelligence, and adapt as technologies, risks, and strategic priorities evolve.

The organisations that succeed in the digital economy will not be those with the most data or the most advanced algorithms, but those that govern intelligence deliberately, invest wisely, and consistently translate insight into outcomes.

Bibliography

Agrawal, A.K., Gans, J.S. & Goldfarb, A. (2018, May). 'Exploring the impact of artificial intelligence: Prediction versus judgement', National Bureau of Economic Research. http://www.nber.org/papers/w24626%20 [© 2018 by Ajay K. Agrawal, Joshua S. Gans, and Avi Goldfarb. All rights reserved. Short sections of text, not to exceed two paragraphs, may be quoted without explicit permission provided that full credit, including © notice, is given to the source].

Amador, J. & Silva, C. (2023, April). 'The impact of ICT adoption on productivity: Evidence from Portuguese firm-level data', Working Paper No. 202307, Banco de Portugal.

Australian Prudential Regulation Authority (APRA), This work is licensed under the Creative Commons Attribution 3.0 Australia Licence (CCBY 3.0). This licence allows you to copy, distribute and adapt this work, provided you attribute the work and do not suggest that APRA endorses you or your work. © Australian Prudential Regulation Authority (APRA)

Bontadini, F., Corrado, C., Iommi, I., & Jona-Lasinio, C. (2024, June 7). 'Data, Intangible Capital, and Productivity*'. National Bureau of Economic Research. https://www.nber.org/system/files/chapters/c14737/c14737.pdf [* 'Paper originally prepared for the NBER/CRIW Conference on Technology, Productivity, and Economic Growth, March 17-18, 2022. The authors are grateful to the OECD Economics Directorate and the European Commission's DG ECFIN for partial financial support of work reported in this paper.']

Booyse, D. & Scheepers, C.B. (2023, June 27). 'Barriers to adopting automated organisational decision-making through the use of artificial intelligence'. *Management Research Review.* https://www.emerald.com/insight/content/doi/10.1108/mrr-09-2021-0701/full/html%20 [© Dawid Booyse and Caren Brenda Scheepers. Published by Emerald Publishing Limited. This article is published under the Creative Commons Attribution (CC BY 4.0) licence].

Carvalho, T. P., Soares, F. A. A. M. N., Vita, R., Francisco, R. da P., Basto, J. P., & Alcalá, S. G. S. (2019). A systematic literature review of machine learning methods applied to predictive maintenance. Computers & Industrial Engineering, 137, 106024. https://doi.org/10.1016/j.cie.2019.106024

Gama, J., Žliobaitė, I., Bifet, A., Pechenizkiy, M., & Bouchachia, A. (2014). A survey on concept drift adaptation. *ACM Computing Surveys*, 46(4), Article 44, 1–37. https://doi.org/10.1145/2523813

Hauser, John & Katz, Gerald. (1998). 'Metrics: you are what you measure!', *European Management Journal*, vol. 16(5): 517-528, October.

Kahneman, D. (2011): *Thinking, fast and slow*, Farrar, Straus and Giroux, New York.

Klein, G.A. (1993). 'A recognition-primed decision (RPD) model of rapid decision making'. In G.A. Klein, J. Orasanu, R. Calderwood, & C. E. Zsambok (Eds.), *Decision making in action: Models and methods* (pp. 138–147). Ablex Publishing. https://psycnet.apa.org/record/1993-97634-006

Levy, F. (2010). 'How technology changes demands for human skills'. (OECD Education Working Paper No. 45). OECD. https://doi.org/10.1787/5kmhds6czqzq-en [The opinions expressed and arguments employed in this adaptation should not be reported as representing the official views of the OECD or of its Member countries.

This is an open-access article distributed under the terms similar to the Creative Commons Attribution License (CC BY 4.0).]

Loftus, E. F. (2005). Planting misinformation in the human mind: A 30-year investigation of the malleability of memory. *Learning & Memory*, 12(4), 361–366. https://doi.org/10.1101/lm.94705

Mills, S., & Spencer, D. A. (2024, December 3). 'Efficient Inefficiency: Organisational challenges of realising economic gains from AI', *Journal of Business Research* 189, 11512, 3 December 2024. https://doi.org/10.1016/j.jbusres.2024.115128 [© 2024 Published by Elsevier Inc. This is an open access article under the CC BY license (http://creativecommons.org/licenses/by/4.0/)].

Mitchell, J., Ker, D. & Lesher, M. (2021). 'Measuring the Economic Value of Data.' Going Digital Toolkit Note No. 20. Paris: OECD, 2021. https://goingdigital.oecd.org/data/notes/No20_ToolkitNote_MeasuringtheValueofData.pdf with an adaptation of an original work by the OECD. The opinions expressed and arguments employed in this adaptation should not be reported as representing the official views of the OECD or of its Member countries. This is an open-access article distributed under the terms similar to the Creative Commons Attribution License (CC BY 4.0).

Mittelstadt, B. D., et al. (2016). 'The Ethics of Algorithms: Mapping the Debate', *Big Data and Society*, vol. 3(2), SAGE Publications, 2016. https://doi.org/10.1177/2053951716679679%20

National Institute of Standards and Technology. (2023). Artificial intelligence risk management framework (AI RMF 1.0) (NIST AI 100-1). U.S. Department of Commerce. https://doi.org/10.6028/NIST.AI.100-1

Parker, M.M. & Benson, R.J. (1988). *Information economics: Linking business performance to information technology*, Englewood Cliffs, New Jersey: Prentice Hall.

Poplin, R., Varadarajan, A. V., Blumer, K., Liu, Y., McConnell, M. V., Corrado, G. S., Peng, L., & Webster, D. R. (2018). Prediction of cardiovascular risk factors from retinal fundus photographs via deep learning. *Nature Biomedical Engineering*, 2(3), 158–164. https://doi.org/10.1038/s41551-018-0195-0

Porter, M.E. (1985). *The Competitive Advantage: Creating and Sustaining Superior Performance*, NY: Free Press.

Romeo, G., Conti, D. (2025). 'Exploring automation bias in human–AI collaboration: a review and implications for explainable AI', *AI & Soc* (2025). https://doi.org/10.1007/s00146-025-02422-7

Rosbach, E., Ganz, J., Ammeling, J., Riener, A., Aubreville, M. (2025). 'Automation Bias in AI-assisted Medical Decision-making under Time Pressure in Computational Pathology', In: Palm, C., et al. *Bildverarbeitung für die Medizin* 2025. BVM 2025. Informatik aktuell. Springer Vieweg, Wiesbaden. https://doi.org/10.1007/978-3-658-47422-5_27

Siddique, S.; Haque, M.A.; George, R.; Gupta, K.D.; Gupta, D.; Faruk, M.J.H. (December 2023): 'Survey on Machine Learning Biases and Mitigation Techniques', *Digital* 2024, 4, 1–68. https://doi.org/10.3390/digital4010001 [© 2023 by the authors. Licensee MDPI, Basel, Switzerland. This article is an open access article distributed under the terms and conditions of the Creative Commons Attribution (CC BY) license (https://creativecommons.org/licenses/by/4.0/)].

TM Forum (2022), 'GB1023 Data Governance Guidebook v3.0.0 (01 June 2022)'. Adapted from the original work of TM Forum and should not be interpreted as representing the official views of TM Forum. Copyright © TM Forum 2022. All Rights Reserved. This document and translations of it may be copied and furnished to others, and derivative works that comment on or otherwise explain it or assist in its implementation may be prepared, copied, published, and distributed, in whole or in part, without restriction of any kind, provided that the above copyright notice and this section are included on all such copies and derivative works.

Topol, E. J. (2019). High-performance medicine: The convergence of human and artificial intelligence. *Nature Medicine*, 25(1), 44–56. https://doi.org/10.1038/s41591-018-0300-7

Tversky, A. & Kahneman, D. (1974). 'Judgement under Uncertainty: Heuristics and Biases', *Science*, 185(4157), 1124–1131.

U.S. Department of Labor, Employment and Training Administration (O*NET 29.3 Database by the USDOL/ETA). https://www.onetcenter.org/database.html Used under the CC BY 4.0 license. O*NET® is a trademark of USDOL/ETA. The authors of this book have modified all or some of this information. USDOL/ETA has not approved, endorsed, or tested these modifications.

Wilson & Daugherty (2018). Collaborative Intelligence: Humans and AI Are Joining Forces. Harvard Business Review

Woolley et al. (2010). Evidence for a collective intelligence factor in groups. Science

Zaimoglu, A.K., Pratt, L. & Fisher, B. (2023, November 23). 'Epistemological role of human reasoning in data-informed decision-making', *Frontiers in Communication*, 23 November 2023. https://doi.org/10.3389/fcomm.2023.1250301%20 [© 2023 Zaimoglu, Pratt & Fisher. This is an open-access article distributed under the terms of the Creative Commons Attribution License (CC BY)].

Subject Index

W

www.ingramcontent.com/pod-product-compliance
Lightning Source LLC
LaVergne TN
LVHW071334210926
848486LV00003B/147
9781764273190